Ubiquitous Quantum Structure

Andrei Khrennikov

Ubiquitous Quantum Structure

From Psychology to Finance

 Springer

Prof. Andrei Khrennikov
University of Växjö
International Center for Mathematical
Modeling in Physics and Cognitive Science
Vejdes Plats 7
SE-351 95 Växjö
Sweden
andrei.khrennikov@vxu.se

ISBN 978-3-642-05100-5   e-ISBN 978-3-642-05101-2
DOI 10.1007/978-3-642-05101-2
Springer Heidelberg Dordrecht London New York

Library of Congress Control Number: 2009943303

Cover design: deblik, Berlin

Printed on acid-free paper

Springer is part of Springer Science+Business Media (www.springer.com)

Preface

The aim of this book is to demonstrate that *quantum-like* (QL) models, i.e., models based on the mathematical formalism of quantum mechanics (QM) and its generalizations, can be successfully applied to *cognitive science, psychology, genetics, the economy, finances, and game theory.*

This book is not about quantum mechanics as a physical theory. The short review of quantum postulates has merely historical value: quantum mechanics is just the first example of successful application of *non-Kolmogorovian probabilities*, the first step towards a *contextual probabilistic* description of natural, biological, psychological, social, economic or financial phenomena. I have developed a general *contextual probabilistic model* (Växjö model) that can be used to describe probabilities in both quantum and classical (statistical) mechanics as well as in the above-mentioned phenomena. This model can be represented in a QL way, namely, in complex and more general Hilbert spaces. In particular, quantum probability is totally demystified: *Born's representation* of quantum probabilities by complex probability amplitudes, wave functions, is simply a special representation of this type. QL representation of data is very convenient; it can be used in any domain of science. I have presented [180, 198] a fundamental conjecture that some biological systems might develop the ability to create QL representations of external and internal worlds. Starting with this conjecture, QL models of cognitive and psychological processes are developed.

A simple statistical test of QL probabilistic behavior based on *interference of probabilities* was elaborated [180, 176] and corresponding experiments have been performed by Conte et al. [66, 67]. The conjecture on the QL processing of information in the brain was confirmed: ensembles of students performing incompatible recognition tasks for recognition of ambiguous figures demonstrated nontrivial interference of probabilities, see Chapter 6.

Recently a professor of cognitive psychology, Jerome Busemeyer, conjectured that probabilistic data obtained in famous experiments on the so-called *disjunction effect* by Tversky and Shafir [295, 275] cannot be described by the conventional Markovian probabilistic model, and he speculated that the disjunction effect can be described by quantum formalism. This viewpoint was elaborated in works by Busemeyer et al. [48–50]. We recall that the disjunction effect is by definition (given by Tversky and Shafir[1]) an exhibition of violation of Savage's *Sure Thing Principle* (STP) [271]. The latter is the basis of the modern theory of *rational decision making*. In particular, by Savage's axiomatics, economic processes are based on actions of rational agents of the market (in particular, of the financial market), i.e., on the STP. In this book I apply the contextual approach to the disjunction effect. A QL model is presented in Chapter 7; see [208, 206, 213, 209] for original publications. It seems that, in spite of Busemeyer's conjecture, the conventional quantum formalism is too restrictive to describe the disjunction effect. It is impossible to describe this effect not only by classical (Kolmogorovian) probability theory, but even by conventional quantum probability (defined by Born's rule in Dirac–von Neumann's formalism of QM). I apply a generalization of the QM-formalism that is naturally generated in the contextual probabilistic framework. My model is based on the assumption that, in the process of evolution, cognitive systems developed the ability to represent contexts by probabilistic amplitudes (complex and even more general). Such amplitudes form a linear space. Thus the brain is able to linearize probabilistic images of contexts. The dynamics is described by a linear evolution equation, *a mental Schödinger's equation.* Consequently, decision making is described mathematically by quantum (and more general QL) theory, see, e.g., Holevo [147, 148], Helstrom [141] or Marley and Hornstein [238]. We also mention that the interest in applications of quantum and QL methods to decision making in cognitive science, psychology and economics is very large, see publications by Danilov and Lambert-Mogilansky [73–76] (utility theory), La Mura [223] (utility theory) (see also [222]), Franco [109–114] (cognitive psychology), and Haven and Khrennikov [137] (a fundamental work covering all possible paradoxes in cognitive psychology related to the disjunction effect).

The crucial point is that QL probabilistic behavior, e.g., in research on brain functioning, need not be a consequence of special physical conditions. For example, the hot brain can still produce interference of probabilities. This is a significant advantage of the QL approach compared with quantum physical reductionism. I have named this approach the *QL paradigm.* A detailed presentation of this paradigm can be found in the first chapter of this book.

This book is truly intended to be accessible to psychologists and researchers working in cognitive science, sociology or economics. Therefore, the first chapter provides a detailed review of all the basic ideas and methods used in the following chapters, without containing any mathematics. It might even be useful for philosophers interested in quantum foundations and the QL description of physical, biological, and mental phenomena.

Växjö, Moscow   *Andrei Khrennikov*
2009

---

[1] See also the excellent experimental work of Croson [71] confirming and generalizing results [295, 275].

Acknowledgements

I thank L. Accardi, D. Aerts, J. Busemeyer, E. Conte, A. Elitzur, A. Ezhov, R. Franco, W. Freudenberg, A. Grib, S. Hameroff, E. Haven, A. Lambert-Mogil/ansky, M. Ohya, P. Suppes, and K. Svozil for fruitful discussions on applications of quantum probability outside of physics. This book was completed during my visits to the University of Lecce and the Centro Vito Volterra of the University of Rome 2, June 2008, and the Steklov Mathematical Institute of the Russian Academy of Science and the Institute of Information Security of the Russian State University for Humanities, September–December 2008. I thank C. Garola, L. Accardi, I. Volovich, S. Kozyrev, and V. Maksimov for hospitality.

Contents

Chapter 1
Quantum-like Paradigm

The *Quantum-like paradigm* (QL) is based on understanding that the mathematical apparatus of quantum mechanics (and especially quantum probability) is not rigidly coupled with *quantum physics*, but can have a wider class of applications.

1.1 Applications of Mathematical Apparatus of QM Outside of Physics

Recall that differential and integral calculi were developed to serve classical Newtonian mechanics. However, nowadays nobody is surprised that these tools are widely used everywhere – in engineering, biology, economics, ... In the same way, although the mathematical apparatus of quantum mechanics was developed to describe phenomena in the microworld, it could be applied to the solution of various problems outside physics.

One of the interesting problems is to apply *quantum probability*, e.g., to cognitive science or to financial markets. The main distinguishing features of quantum probability is its representation by the *complex probability amplitude*. In the abstract approach, such an amplitude is represented by a normalized vector in a complex Hilbert space, while the so-called mixed state is represented by a *density matrix*. Probability (which is compared with experimental relative frequencies) is given by *Born's rule*. For example, for measurement of the coordinate x of a quantum particle, the probability of finding it at the point $x = x_0$ is equal to the square of the absolute value of the wave function at this point.

The main (merely psychological) barrier in applications of QL models outside quantum physics is a rather common opinion that the "unusualness" of the quantum formalism (compared with, e.g., classical statistical mechanics) is an exhibition of "unusualness" of quantum systems. Such rather mystical things as *quantum non-locality* and *death of realism* are firmly coupled to the modern interpretation of quantum mechanics (in particular, through Bell's theorem and other "no-go" theorems). It seems that quantum formalism cannot be applied to "usual systems" (e.g., macroscopic biological systems or huge social systems), because in this case it is not

A. Khrennikov, *Ubiquitous Quantum Structure*,
DOI 10.1007/978-3-642-05101-2_1, © Springer-Verlag Berlin Heidelberg 2010

easy to accept death of realism[1] or any sort of nonlocality. For instance, attempts to describe the brain's functioning by means of QM typically induce heavy discussions on time, space, and temperature scales. The aim of such scale studies is to couple the brain's scales to quantum scales.

My aim is to show that such a reduction to quantum physical scales is totally unnecessary. The hot macroscopic brain might be able to process information in a quantum-like way, exhibiting, e.g., the interference of probabilities of alternatives. Moreover, any sufficiently complex biological, social, or financial system might exhibit quantum-like probabilistic features. I shall explain the source of such features a little bit later after the discussion on quantum randomness and probability. At the moment I just point out *contextuality* as the main source of quantum-like probabilistic behavior.

1.2 Irreducible Quantum Randomness, Copenhagen Interpretation

During the past 70 years the development of quantum mechanics has been characterized by intense debates on the origin of quantum randomness and in particular on possibilities to reduce it to the classical ensemble randomness. For example, von Neumann was convinced that *quantum randomness is irreducible,* but Einstein had the opposite view on this problem: for him the discovery of quantum mechanics was merely a discovery of a special mathematical formalism (quantum formalism) for description of a special *incomplete representation of information* about microsystems.

According to the Copenhagen interpretation of QM a pure quantum state (wave function) describes an *individual quantum system*, not an ensemble of systems in the sense of classical probability. As a consequence of such an "individual interpretation", a concrete physical system, e.g., an electron, can be prepared in a physical superposition of pure states.

In the majority of textbooks on QM we can read about, e.g., an atom in a superposition of different energy states[2], or an electron in a superposition of spin-up and spin-down states; in the famous two-slit experiment a photon is in a superposition of passing through both slits.

An attempt to apply the mathematical formalism of QM outside of the microworld in combination with the Copenhagen interpretation would create obvious difficulties: it is not easy to imagine a macroscopic system, e.g., in economics, that is in a real, physical, superposition of two states. Of course, I am well aware of the existence of macroscopic quantum systems as well as of the attempts to use the Copenhagen interpretation even in this case – e.g., by Legget in superconductivity,

---

[1] For example, neurons in the brain definitely have objective properties.

[2] Unlike in QM, in classical mechanics the energy of a particle can (at least in principle) be exactly determined.

by Zeilinger in the two-slit experiments for macroscopic systems, by De Martini in experiments with "macroscopic Schrödinger cats" etc.

It is well known that such attempts to proceed with the Copenhagen interpretation for macroscopic quantum systems do not provide a clear physical picture of the phenomena. One of the possibilities is to use *de Broglie's wavelength* for characterization of the wave features of a macroscopic system. Since it is very small for a large system, it is always possible to say that, although a macroscopic system has wave features, they are hardly observable.

This kind of compromise is hardly satisfactory as a solution to a conceptual problem, especially for justification of applications outside quantum physics, e.g., in psychology or economics. Moreover, as already pointed out by Pauli in the early times of QM, any attempt to interpret the wave function as a physical wave clashes against the fact that, for most interesting physically systems, these wave functions are defined in a multidimensional mathematical space.[3]

Thus the supporters of the *wave–particle duality* face the paradox of believing in *a physical wave in a nonphysical space* (already in the case of a two-particle system).

1.3 Quantum Reductionism in Biology and Cognitive Science

As a consequence of the above-mentioned difficulties with the interpretation of macroscopic quantum systems, a popular attitude today in attempts to apply quantum mechanics (e.g., in biology) is to proceed beyond conventional (e.g., biological) models that operate with states of macroscopic systems.

For example, in cognitive science a group of researchers (e.g., Penrose [249, 250] and Hameroff [128, 129]) developed the reductionist approach to the brain's functioning. They moved beyond the *conventional neuronal paradigm* of cognitive science and tried to reduce processing of information in the brain to quantum microprocesses – on the level of quantum particles composing the brain. Penrose repeated many times that a neuron (as a macroscopic system) could not be in a physical superposition of two states: firing and nonfiring.

As was already mentioned, the majority of attempts to apply the mathematical formalism of quantum mechanics outside physics were based on the reduction of the processes under consideration to some underlying quantum processes in the microworld. This reductionist approach was heavily based on the following argument: since everything in this world is composed of quantum particles, any kind of process might be (at least in principle) reduced to a quantum process.

---

[3] The wave function of, e.g., a pair of electrons is defined not on physical space described mathematically by the cartesian product of 3 real lines, but on configuration space, which is the cartesian product of 6 real lines. Thus, Schrödinger already understood well that two electrons cannot be embedded in physical space. Therefore he gave up with his interpretation of the wave function as charge density. In principle, already at this stage one might start to speak about "quantum nonlocality", i.e., without any reference to Bell's inequality.

The unification dream is in principle correct, and it has played an important role in the development of natural sciences; in this spirit any attempt to apply quantum mechanics to, e.g., cognitive science should be welcome. However, it is very difficult (if possible at all) to establish a natural correspondence between conventional macroscopic models and underlying quantum models.[4] There is a huge difference in scales of parameters in those models. Moreover, even in quantum physics the *correspondence principle* is vaguely formulated and not totally justified and, on the other hand, even in classical physics, the unification dream is far from being accomplished in spite of the important successes of statistical mechanics in the reduction of thermodynamics to mechanics. For example, structures such as crystals, which are relatively simple in comparison with biological structures, at the moment have not been deduced from first principles in either classical or quantum physics.

1.4 Statistical (or Ensemble) Interpretation of QM

I point out that it is possible to escape the above-mentioned difficulties by rejection of the Copenhagen interpretation and association of a pure quantum state (wave function) not with an individual quantum system, but with an ensemble of systems.

Such an interpretation is called the *statistical (or ensemble) interpretation* of QM. It was originally proposed by many authors, including Einstein, Popper, Margenau, de Broglie, Bohm, and Ballentine, but only with the development of quantum probability could it overcome the traditional criticism which prevented, for over 50 years, the majority of physicists from accepting this apparently natural interpretation. The main objection to it, to which the above-mentioned authors never gave a satisfactory answer, was that the statistical interpretation is contradicted by the experimental data. We recall that at the beginning Schrödinger was quite sympathetic to Einstein's attempts to proceed in the quantum framework on the basis of the statistical interpretation.[5]

Schrödinger wrote that if Einstein were able to derive *interference of probabilities* for the two-slit experiment on the basis of the statistical model, he (Schrödinger)

---

[4] This whole "quantum approach" is very speculative because it is currently controversial whether quantum theoretical mechanisms can be experimentally identified in the neural correlates (or constituents) of cognitive processes such as, e.g., decision-making.

[5] It is practically forgotten that the famous *Schrödinger cat* was created to show the absurdness of the Copenhagen interpretation of QM. If one accepts superposition of states for an individual microscopic system, then superposition of states for an individual macroscopic system should also be accepted. We recall that Schrödinger just modified Einstein's example with a gun and a man by making it more peaceful (at least at that time) – by using poison and a cat. We remark that nowadays people (heavily infected by the Copenhagen interpretation) take Schrödinger cats seriously. A number of famous experimental groups produce Schrödinger cats (as they believe) in their labs. Of course, I have no doubt that such experiments as, e.g., done by the group of De Martini from the University of Rome ("La Sapienza") are great contributions in the domain of quantum foundations, but the belief that really Schrödinger cat-type states are produced is rather naive.

would definitely choose this model. However, neither Einstein nor anybody else was able to perform such a derivation.

Concerning this objection, my main point is that the experimental data *contradict the use of the Kolmogorov model* of probability and not the statistical interpretation by itself. If one keeps to the statistical interpretation, then one can assume that the quantum probabilistic description need not be based on irreducible quantum randomness. However, the classical probability model should be generalized to take into account *contextuality* of probability, i.e., its dependence on context (physical, biological, mental, social, financial) of observations (which could even be self-observations of, e.g., the brain).

The contextual probabilistic calculus can be used for an incomplete description of statistical data. One could not even exclude that in some cases a Kolmogorov model can be found beyond the QL contextual probabilistic description. The crucial point is that the role of the presence of a "hidden Kolmogorovian model" is negligible if one has no access to data described by the latter (typically unobservable joint probabilities). In such cases the only reasonable possibility is to use the quantum-like probabilistic description or different non-Kolmogorovian models.

Thus we propose testing in various applications the approach based on accepting Einstein's viewpoint: the mathematical formalism that was developed to serve quantum physics is a special form of *incomplete probabilistic description*. Of course, for QM (as a physical theory) Einstein's viewpoint implies its incompleteness.

1.5 No-Go Theorems

The natural question which is typically asked as the first reaction to my proposal is the following:

What about the known no-go theorems?

I will not enter here into a debate on the complicated problem of the validity of no-go theorems.[6] In fact, the main problem of the "no-go ideology" is that it is directed against *all possible* prequantum models (the so-called hidden variable models). Supporters of no-go activity formulate new theorems excluding various classes of models with hidden variables, but one can never be sure that a natural model that does not contradict any known no-go theorem will finally be found.

Remark 1.1 I do not agree with Bell's attempt to couple the so-called "quantum nonlocality" with the problem of completeness of quantum mechanics. Einstein, Podolsky and Rosen [99] considered "quantum nonlocality" as an absurd alternative to incompleteness. Unfortunately, nowadays quantum nonlocality has become extremely popular in quantum information theory. Moreover, this idea is diffusing

---

[6] See, for example, my books [159, 161, 214] and papers [162, 164] as well as my papers with Igor Volovich [166] and Luigi Accardi [4]; see also stormy debates in the proceedings of Växjö conferences [165, 167, 5, 6].

outside quantum physics: it has become fashionable to refer to quantum nonlocality in cognitive and social sciences and even in parapsychology. In the latter case quantum nonlocality provides really great new possibilities. Conferences devoted to Quantum Mind have become a tribune for parapsychological speculations based on quantum nonlocality. Of course, people working in quantum information theory and trying to design quantum computers, cryptography and teleportation are not so happy to hold joint meetings with, e.g., "quantum buddhists" creating a powerful new religion, but they have no choice! By accepting "quantum nonlocality" they are in one camp with people providing the QM-interpretation for a nonlocal deity.

As a sign of inconsistency of the no-go activity, we mention the sharp criticism of the assumptions of known no-go theorems by newcomers – authors proposing new no-go statements. For instance, Bell criticized [31] quite aggressively assumptions of von Neumann's no-go theorem [301] (and other no-go theorems which existed at that time). Therefore it is surprising that nowadays the majority of the quantum community (especially people working in quantum information) reacts so painfully to critique of the assumptions of Bell's theorem – for such a critique see, e.g., Accardi [3], Accardi and Khrennikov [4], Adenier and Khrennikov [7], Andreev and Man'ko [17], De Baere [78, 79], De Muynck et al. [84, 85], Hess and Philipp [142, 143], Khrennikov [159, 161, 162, 164, 179], Khrennikov and Volovich [166, 193], Klyshko [216, 217]. We point out especially the practically forgotten papers by Klyshko. He was one of the best experimenters in the world in the domain of quantum optics. It is amazing that he came to the same conclusion as pure mathematicians, e.g., Accardi, Aerts, Khrennkiov, and recently Hess and Philipp.

Violation of Bell's inequality is merely an exhibition of non-Kolmogorovness of quantum probability, i.e., the impossibility of representing all quantum correlations as correlations with respect to a single Kolmogorov probability space [219]; there is no direct relation between this violation and such mysterious things as nonlocality and death of realism.

1.6 Einstein's and Bohr's Views on Realism

I emphasize that Einstein's realism is a quite naive form of realism. It does not take into account the dynamics of the process of interaction of a system with the measurement device. The father of the Copenhagen interpretation N. Bohr permanently pointed out that the *whole experimental arrangement* (context) should be taken into account. It is the crucial point not only for physics, but for other domains of science. For example, in the process of decision making the brain interacts with questions (problems). The resulting decision is the result of such an interaction.

I share Einstein's views only partially. I keep to the statistical (ensemble) interpretation of the quantum state and consequent incompleteness of QM, but I agree with Bohr in considering the values of some quantum observables as responses to interactions with apparatus rather than objective properties of quantum systems. The

latter point has been emphasized by Luigi Accardi since the early 1980s in a series
of works on the so-called *chameleon effect*, see [1–4]. This effect is nothing other
than the dependence of a chameleon's color (its "property") on a surface's color.
The latter is an analog of the state of apparatus. Of course, such a type of system–
apparatus interaction is purely deterministic. To be closer to reality, one should
consider random chameleons. Another similar approach is the *adaptive dynamics*
developed by Masanori Ohya, see, e.g., [245, 246].

On the basis of such an ideology one can proceed successfully.

This combination of the views of Einstein and Bohr is known as the *Växjö interpretation* of QM, see [177].

Concerning no-go theorems my proposal could be summarized as follows:

*"Do not be afraid to consider the quantum description as an incomplete one.
Look for applications of quantum formalism outside quantum physics!"*.

1.7 Quantum and Quantum-like Models

As a comment on the use of the notion quantum-like (QL) behavior, I think that it
would be useful to preserve the term "quantum" for quantum physics while, in other
models which are still based on quantum or, more generally, non-Kolmogorovian,
probabilistic description, we should use the term "quantum-like". In particular, in
this way my approach can be distinguished from a purely reductionist one. For
example, the quantum brain model is a reductionist model of the brain functioning,
but the *QL brain* model is a model in which the wave function provides a (incomplete) probabilistic representation of information produced by the neurons and not
a model of the actual physical state of them. In the same way a *quantum game* is
based on randomness produced by quantum physical systems (e.g., photons), but
a *QL game* can be performed by purely classical physical systems (e.g., people)
exhibiting QL probabilistic behavior.

1.8 Quantum-like Representation Algorithm – QLRA

Quantum-like modeling immediately meets one complex problem: the creation of
QL-representations (in complex and more general Hilbert spaces) of classical (contextual) probabilistic data. For example, looking for a QL model of the brain's functioning we should be able to answer the following question:

*How does the brain represent statistical information by the wave function (com-
plex probability amplitude)?*

If one considers the brain as a kind of probabilistic machine, then this problem
can be formulated as the *inverse Born problem*:

To construct a complex probability amplitude (or in the abstract framework a normalized vector of Hilbert space) on the basis of probabilistic data. This amplitude should produce probabilities by Born's rule.

An attempt to solve this problem was made in a series of my works, starting with [163]. There was created the so-called *QL representation algorithm – QLRA.*[7] It transforms probabilistic data of any origin (which should satisfy some natural restrictions) into a complex probability amplitude.

1.9 Non-Kolmogorov Probability

We now couple the QL paradigm with another important probabilistic paradigm. Nonclassical statistical data are not covered completely by the conventional quantum model. The main distinguishing feature of quantum probability is its *non-Kolmogorovinity* expressed in the form of contextuality.[8]

It was emphasized (by Luigi Accardi, Diederik Aerts, Stan Gudder and me[9]) that in the same way as in geometry (where, starting with Lobachevsky, Gauss, Riemann, ..., various non-Euclidean geometries were developed and widely applied, e.g., in relativity theory), in probability theory various non-Kolmogorov models may be developed to serve applications. The QM probabilistic model was one of the first non-Kolmogorovian models that had important applications. Thus one may expect development of other types of probabilistic models which would be neither Kolmogorovian nor quantum.

---

[7] Improvement of this algorithm, its generalization, and creation of new QL representation algorithms is an important problem in the realization of the QL paradigm.

[8] We remind the reader that the main distinguishing feature of the Kolmogorov [219], 1933, model of probability is the possibility of embedding everything (probabilities and obsevables – called random variables) in a single space Ω. In particular, all probabilities are reduced from a single probability measure on this space, say **P**, and all random variables are realized by functions on this space. We also point out that, in spite of rather common opinion, the Kolmogorov model is not so simple mathematically. The measure-theoretic considerations are complicated – much more complicated than the linear algebra of quantum probability. In fact, the main difficulties in quantum probability are related to the infinite dimension of the state space, Hilbert space. However, in many applications, e.g., quantum information, one can proceed with linear algebra in finite-dimensional spaces.

[9] There are some debates on priority. However, I think that such debates are totally meaningless. These are debates about the first two bright stars in the complete darkness of the traditional quantum kingdom. I was strongly influenced by all of them. Conversations with Stan Gudder and Luigi Accardi during their visits to Växjö played an important role in the formation of my views on nonclassical probability. Although I did not succeed in inviting Diederik Aerts to one of the conferences of the Växjö series on foundations of quantum theory and quantum information, e-mail exchange with him also was very important for me.

1.10 Contextual Probabilistic Model – Växjö Model

In this book a general scheme of probabilistic description of experimental probabilistic data is presented. We call this model the contextual probabilistic model (the *Växjö model*) [185]. The origin of the data does not play any role. It could be statistical data from quantum mechanics or equally from classical statistical mechanics, biology, sociology, economics, or meteorology. Then it will be shown that the interference of probabilities (even more generally than in quantum mechanics) can be found for any kind of data [163, 170–174, 176–178, 186–188]. As was pointed out, one can easily obtain the linear space representation of probabilities from the interference of probabilities (by applying QLRA) and then recover Born's rule (which will not be a postulate anymore). Thus, in our approach the quantum probabilistic calculus is just a special linear space representation of given probabilistic data [163]. One of the advantages of the QL representation of probabilistic data is an essential simplification of operating with this data – *linearization* of a model always induces simplification.

Basic to our approach is the notion of *context* – a complex of conditions under which the measurement is performed. Contexts of different kinds can be considered: physical, biological or even political. Our approach to the subject of probability is contextual. It is meaningless to consider a probability not specifying the context of consideration.

Kolmogorov was well aware of the *contextuality* of his probability space [219]. He emphasized that any experiment should be described by its own (Kolmogorov) probability space. Unfortunately, this ideological recognition of contextuality of probabilities did not imply any form of the mathematical formalization in his work (or in the work of later users of the Kolmogorov probability model). One of the sources of quite misleading (for applications) development of the mathematical theory of probability was the presence of conditioning in the standard probability model. Kolmogorov defined conditional probabilities by using Bayes formula. It became a custom to identify context-dependence with such *Bayesian conditioning*. I think that it restricted essentially the range of context-dependent phenomena. Already in QM, one should go beyond Bayesian conditioning. It is amazing that QM was created at the same time as the Kolmogorovian model. But mathematicians were able to proceed for at least 50 years without understanding that this model suffers from a very special definition of conditioning.

I should also mention the model of Hungarian mathematician Renyi [267], which is nowadays practically forgotten. He tried to extend Bayes-Kolmogorov conditioning. But the closest to my approach were the probabilistic studies by Mackey [235], who tried to reconstruct quantum probability on the basis of contextual (conditional) probabilities. Unfortunately, his attempt was not completely successful. Finally, he postulated the Hilbert space representation of probabilities.

As was pointed out, generally the QL paradigm is not reduced to application of conventional quantum probability (based on the complex Hilbert space representation) outside quantum physics. Other (non-Kolmogorovian) models might be applied as well. One such model was presented – it is the model with so-called *hyperbolic interference* of probabilities [163]. It is based on representation of probabilities by amplitudes taking values in the algebra of so-called hyperbolic numbers (a two-dimensional Clifford algebra).[10]

1.11 Experimental Verification

Finally, one should analyze several different families of empirical data in order to realize concretely the outlined programme. It is clear that without experimental justification the QL paradigm is simply a philosophical-mathematical principle. It can give rise to a huge variety of interesting mathematical models (as was done, e.g., in string theory [123]), but finally these models should be verified by comparison with experimental statistical data.

The first steps of the experimental verification have already been done, see Conte et al. [66, 67]. It was experimentally confirmed that the brain behaves as a QL system in the process of decision making (in tests with ambiguous figures): the interference of probabilities related to incompatible questions was found. Moreover, it was natural to suppose that some well-known experiments that have already been done in cognitive science or psychology might produce nonclassical statistical data.[11] Recently Busemeyer et al. [48, 49] pointed out that statistical data collected in famous experiments in cognitive psychology performed by Shafir and Tversky [275, 295] do not agree with the standard Markov chain model based on classical probability theory. He proposed using the quantum model. Stimulated by discussions with Jerome Busemeyer (professor of cognitive psychology), I applied my QL approach to Shafir-Tversky statistical data. It was represented (via QLRA) by complex (and in some case more general hyperbolic!) probability amplitudes. The interference of probabilities was found [208, 213, 206], see also the joint work with Emmanuel Haven [207]. I remark that, opposite to the original conjecture of Busemeyer, Shafir-Tversky statistical data do not match the conventional quantum model. The corresponding matrices of *transition probabilities* are not *doubly stochastic*, but they should be by QM (in the case of observers with nondegenerate spectra).

---

[10] This example motivates extension of the QL paradigm by attempting to develop and apply models in which probability amplitudes take values in various commutative and even noncommutative algebras. The corresponding generalizations of Born's rule should be presented, analogues of the QL representation algorithms should be created. These are interesting and complex problems!

[11] Luigi Accardi pointed out such a possibility many years ago (during my visit to Centro V. Volterra, University of Rome - 2, in 1994). We suspected that nonclassical data might be found in economics, finances, and psychology. The same point was expressed in long discussions with Stan Gudder and Karl Gustafson during my visit to the Universities of Denver and Boulder, in 2001. However, it was really impossible to find such data without competence in the corresponding domains of science. Therefore I was happy when Elio Conte proposed performing experiments to test my QL model [180] of cognition.

1.12 Violation of Savage's Sure Thing Principle

We recall that experimental studies by Shafir and Tversky [275, 295] were about behavior of people in games such as the well-known Prisoner's Dilemma, which illustrate Savage's *Sure Thing Principle* (STP) [271]. Violation of this principle is known as the *disjunction effect*, see Shafir and Tversky [275, 295] and also Rapoport [266], Hofstadter [145, 146] and Croson [71]. We recall that STP is one of the basic principles of modern theoretical economics. In fact, it is a form of the postulate on *rational behavior* of agents acting at the market. Thus experimental violation of STP (disjunction effect) is a sign of *irrational behavior* of agents.

The QL model, see Chapter 7, explains this irrationality. The QL extension of probabilistic description of natural and social phenomena shows that the notion of rationality is itself dependent on a probabilistic model. The conventional rationality of economic agents is in fact the classical probabilistic rationality. If the brain processes information by using some QL representation of probabilistic data, then there are no reasons to expect exhibitions of conventional rationality. Of course, such cognitive systems behave irrationally from the conventional (say Kolmogorovian) viewpoint. However, they are completely rational in the corresponding QL sense.

1.13 Quantum-like Description of the Financial Market

In the conventional financial models, rationality of agents of the financial market is formalized through the *efficient market hypothesis*, which was formulated in the 1960s, see Samuelson [269, 270] and Fama [103] for details:

A market is said to be efficient in the determination of the most rational price if all the available information is instantly processed when it reaches the market and it is immediately reflected in a new value of prices of the assets traded.

From the viewpoint of the QL paradigm the most rational price in the sense of conventional theory of the financial market is the most rational in the framework of the classical probabilistic model (the Kolmogorovian measure-theoretic model).[12] If agents of the financial market behave nonclassically (and that is my conjecture), then they use another type of rationality, the QL rationality. QL-rational behavior may look irrational in the conventional framework. (We remark that the experiments of Shafir and Tversky [275, 295] were done to show the irrationality of agents.) We recall once again that QL models describe the following situation. For each context *C*, only a special part of (statistical) information on this context can be measured and hence be available, for example, to traders. And we emphasize once again that such a viewpoint on the conventional quantum model contradicts the Copenhagen interpretation. However, the latter does not disturb us much. Debates on hidden

---

[12] "All the available information" has Boolean structure. The corresponding probability is described by the Kolmogorovian model.

variables in QM may continue for the next few hundred years. Even if physical quantum randomness is irreducible, one cannot exclude the possibility that quantum and more general QL formalisms can describe reducible (but contextual) randomness in other domains of science, for example, psychology or economics.

In Chapter 10 of this book we shall present a QL model of the functioning of agents of the financial market which is based on the mathematical formalism of Bohmian mechanics.[13] The "financial pilot wave" describes expectations of agents. Nonlocality of this model is purely classical: the common field of expectations is created through classical communication channels (TV, internet, newspapers, private communications). The presence of such a field provides a possibility to "rule" the financial market – in the same way as one can rule quantum particles by manipulating the pilot wave.

We remark that the original Bohmian model of QM is totally deterministic. Each quantum particle has well-defined position and momentum. Motion is described by a kind of Newton's equation (say *Bohm-Newton equation*), which contains not only the classical force generated by the physical potential V (for example, the Coulomb potential describing interaction of two charged particles, e.g., electron and proton), but also an additional force – the so-called *quantum force*. The latter is produced by the pilot wave, which is described by Schrödinger's equation (so mathematically it is just the wave function). It can be changed by modification of the physical potential V (which can depend on time) in Schrödinger's equation. The main point is that a slight modification of V can produce a modification of the pilot wave such that the corresponding *quantum force* will be essentially different (from that expected for non-modified V). Applying this formalism to the financial market, we see that by slight modification of the financial potential V one can totally change financial forces. Of course, it should be the right modification.[14]

The original Bohmian model can be essentially improved by adding a classical stochastic term to the Bohm-Newton equation describing motion of the quantum particle. This is the Vigier-Bohm model [40]. Such an improvement is especially useful for description of financial processes. The financial version of the original Bohmian model produces price trajectories that are solutions of the ordinary differential equation (*Bohm-Newton equation*) that is "controlled" by the field of expectations, the pilot wave. The latter makes the model essentially more realistic than classical financial models. Nevertheless, the presence of deterministic price trajectories might be considered a weakness of the Bohmian financial model.[15]

---

[13] This model was invented by the author and Olga Choustova [53, 62, 175] and it was generalized by Emmanuel Haven [136] who made it closer to applications.

[14] Development of the present financial crisis might be considered as indirect confirmation of our model. The effects of special manipulations in the mass media can be considered as modifications of V. Such modifications need not involve many financial resources, so they are really small from the financial point of view.

[15] The problem of whether the financial market is deterministic (a huge deterministic dynamical system, maybe chaotic) or random is still the subject of debate, see the introduction to Chapter 10

The Vigier-Bohm model for the financial market was developed by Olga Choustova [60, 62] and Emmanuel Haven [136].

In fact, the financial Bohmian model is an attempt to go beyond the QL description. One can also proceed with the standard mathematical formalism of QM: Segal and Segal [274], Haven [132–135], Piotrowski et al. [252–258], Danilov and Lambert-Mogiliansky [73, 74], Khrennikov [176]. The model of the financial market based on the use of the Hilbert space representation of financial quantities developed by Baaquie [22] can also be mentioned. The quantum ideology does not play a large role in *Baaquie's model*. It is closer to classical models of *signal analysis* based on the Hilbert space representation.

Finally, we point to the book by Soros [282], which was definitely one of the first works on applications of methods of QM to the financial market. It is amazing that Soros, who was so far from quantum science, claimed that the financial market is a huge quantum system and not at all a collection of classical random systems (as claimed by conventional financial science). It is well known that Soros senses the financial market very well. His heuristic justification of its quantumness is a strong argument in favor of the QL approach. In particular, his book was the starting point for Olga Choustova's and my research.

1.14 Quantum and Quantum-like Games

Game theory plays an important role in various models of economics and evolution theory (including genetic evolution). *Quantum games* naturally arose in the process of development of quantum information, see, e.g., Ekert [100]. In principle, quantum games might be used for modeling of evolution and even in economics. However, one meets the same problems as in attempts to proceed with quantum reductionism in cognitive modeling. Maybe in genetics one could still appeal to conventional quantum games (induced by quantum systems), but in higher level cognitive processes or economics attempts to use quantum games (based on the irreducible randomness of quantum physical systems) are not sufficiently justified. In complete accordance with the QL paradigm, the QL probabilistic behavior can be exhibited in games that are not coupled directly to quantum physical systems – in particular, by (macroscopic) cognitive systems. Thus biological organisms and populations are able to play QL games in the process of evolution. Moreover, the evolution of social and economic systems might be based on QL games.[16] In particular, agents of the

---

for details. Although nowadays various models based on (classical) randomness dominate in theoretical finances, a large amount in real finances is done by technical analysis.

[16] By the QL paradigm the main reason for such games is operating with incomplete information. It can be profitable for a population not to spend too many resources on obtaining complete information about context, but to proceed by operating with incomplete information. Of course, such "nonclassical" operating should be consistent. Some rules should be established. By the QL paradigm various QL rules can be elaborated in this way.

market can play QL games, compare the experimental studies of Shafir and Tversky [275, 295]; see also publications by Piotrowski et al. [252–258].

Even in genetics the quantum-likeness (and not directly quantumness) may be crucial for evolution. Genes and genomes are macroscopic structures. In principle, their functioning can be based on incomplete information processing. In such a case the right model of genetic evolution would be based on a QL game between genes as well as between genomes. However, the latter conjecture is rather speculative. The corresponding experimental studies of quantum-likeness of genetic processes should be performed. On the other hand, simple QL games between people exhibiting interference effects (and even violating Bell's inequality) were proposed by Grib et al. [124, 125] and me [203]; see also the recent paper by Aerts et al. [9].

1.15 Terminology: Context, Contextual Probability, Contextuality

A few remarks regarding the terminology in this book are called for. The notion of the context can be related to the notion of the *preparation procedure*, which is widely used in quantum measurement theory [234, 46, 148]. Of course, preparation procedures – devices preparing systems for subsequent measurements – give a wide class of contexts. However, the context is a more general concept. For example, we can develop models operating with social, political or historical contexts, e.g., socialism context, victorian context. To give another example, one can consider the context "Leo Tolstoy" in literature. The latter context can be represented by various kinds of physical and mental systems – by books, readers, movies.

Contextual probability can be coupled to a *conditional probability*. However, once again the direct identification can be rather misleading, since the conventional meaning of the conditional probability $\mathbf{P}(B|A)$ is the probability that *event B* occurs under the condition that *event A* has occurred [219]. Thus, conventional conditioning is *event-conditioning*. Our conditioning is a *context-conditioning*: $\mathbf{P}(b = \beta|C)$ is the probability that observable b takes the value β in the process of measurement under context C. In principle, we are not against the term "conditional probability" if it is used in the contextual sense.

The main terminological problem is related to the notion of the *contextuality*. The use of the term "contextual" is characterized by a huge diversity of meanings, see Bell [31], Svozil [289] or Beltrametti and Cassinelli [32] for the notion of contextuality in quantum physics as well as Light and Butterworth [228] and Bernasconi and Gustafson [35] for the notion of contextuality in cognitive science and artifical intelligence (AI). In quantum physics the contextuality is typically reduced to a rather specific contextuality – "Bell contextuality." Bell invented this notion in the framework of the EPR-Bohm experiment [31, 32]. We recall that such *quantum contextuality* ("Bell's contextuality") is defined as follows:

The result of the measurement of an observable a depends on another measurement on observable b, although these two observables commute with each other.

It should be emphasized that *nonlocality* in the framework of the EPR-Bohm experiment is a special case of quantum contextuality. Our contextuality is essentially more general than Bell's. In a very special case one can determine the context for the measurement of a by fixing an observable b that is compatible with a. However, in the general case there is nothing about a mutual dependence or compatibility of observables. The context is simply a complex of conditions (e.g. physical or biological). Our description of the EPR-Bohm experiment is contextual, but there is no direct coupling with nonlocality. Our approach to the contextuality is closer to the one used in cognitive science and AI, see [228, 35].

1.16 Formula of Total Probability

The basis of linear representations of probabilities (performed by QLRA) is a generalization of the well-known formula of *total probability* (FTP) [280]. We recall that in the case of two dichotomous variables $a = \alpha_1, \alpha_2$ and $b = \beta_1, \beta_2$ this basic formula has the form

$$\mathbf{P}(b = \beta) = \mathbf{P}(a = \alpha_1)\,\mathbf{P}(b = \beta|a = \alpha_1) \; + \; \mathbf{P}(a = \alpha_2)\,\mathbf{P}(b = \beta|a = \alpha_2),$$

$$(1.1)$$

where $b = \beta_1$ or $b = \beta_2$.

This formula is widely used in statistics and especially in statistical decision making. One wants to predict probabilities for values of the b-variable on the basis of probabilities for values of the a-variable and conditional probabilities to get the value $b = \beta$ under the assumption that $a = \alpha$. In decision making one makes decisions depending on magnitudes of probabilities $\mathbf{P}(b = \beta)$ given by FTP (1.1). Probabilities on the right hand side of FTP could have various interpretations. They could be *objective probabilities* calculated on the basis of the previous statistical experience. They also could be *subjective probabilities* assigned to, e.g., values of the a-variable. Further considerations do not depend on interpretation of probabilities.

1.17 Formula of Total Probability with Interference Term

Starting with the contextual statistical model (Växjö model) we will obtain a generalization of the conventional FTP (1.1) that is characterized by the appearance of an additional term, an *interference term*

$$\mathbf{P}(b = \beta) = \mathbf{P}(a = \alpha_1)\,\mathbf{P}(b = \beta|a = \alpha_1) \; + \; \mathbf{P}(a = \alpha_2)\,\mathbf{P}(b = \beta|a = \alpha_2)$$
$$+ \, \delta(b = \beta|a).$$

$$(1.2)$$

Depending on the magnitude of this term (relative to the magnitudes of probabilities on the right-hand side of (1.1)), we obtain either the conventional *trigonometric interference* ("cos-interference"), which is well known in classical wave mechanics as well as in quantum mechanics, or a *hyperbolic interference* ("cosh-interference"), which was not predicted by conventional physical theories, neither by classical wave theory nor by quantum mechanics. Such a new type of interference arises naturally in the Växjö model.

The possibility of violating the FTP is one of the important consequences of non-Kolmogorovness of probabilistic data. One of the main consequences of QL modeling is that cognitive and social systems can process probabilistic data violating FTP. Such a violation can be negligible for some contexts (which is why conventional FTP has been applied so successfully in many domains of science), but in general it cannot be completely neglected. By neglecting the interference term one comes to paradoxical conclusions, as in the case of data from the experiments of Shafir and Tversky [275, 295]. It is natural to suppose that cognitive and social systems should take into account (to survive in the process of evolution) the mentioned possibility of violation of FTP. Thus they should develop the ability to use a more general probabilistic model than the classical model. We speculate that they use special representations of the contextual statistical model. They may be able to apply QLRA and to represent contextual probabilities in complex or more general linear spaces.

1.18 Quantum-like Representation of Contexts

We recall once again that all probabilities in (1.2) are contextual. They depend on a complex of conditions, context C, for measurements of observables a and b. Starting with FTP with the interference term (1.1) and applying QLRA we obtain two basic types of representations of contexts, $C \to \psi_C$, in linear spaces:

a) representation of some special collection of contexts C^{tr} ("trigonometric contexts")[17] in complex Hilbert space, see Chap. 2;
b) representation of some special collection of contexts C^{hyp} ("hyperbolic contexts")[18] in the so-called hyperbolic Hilbert space.

The complex and hyperbolic representations can be combined in a single representation over a little bit more complicated algebraic structure – the algebra of complex hyperbolic numbers.

We emphasize that in general the collections of trigonometric and hyperbolic contexts, C^{tr} and C^{hyp}, are just proper subsets of the complete collection of contexts C of a Växjö model M (contextual statistical model). Depending on model M there

---

[17] They produce the ordinary cos-interference.

[18] They produce hyperbolic cosh-interference.

can exist contexts that cannot be represented algebraically: neither in complex nor in hyperbolic Hilbert spaces.

We can speculate that some cognitive and social systems might restrict information processing by taking into account only trigonometric contexts. Such systems would process probabilistic information through its representation in the complex Hilbert space.

Chapter 2
Classical (Kolmogorovian) and Quantum (Born) Probability

This chapter contains short introductions to classical and quantum probabilistic models. To simplify presentation, in both cases we consider only *discrete variables*.

2.1 Kolmogorovian Probabilistic Model

We start with two notations. Let A be a set. The *characteristic function* I_A of the set A is defined as $I_A(x) = 1$, $x \in A$, and $I_A(x) = 0$, $x \notin A$. Let $A = \{a_1, \ldots, a_n\}$ be a finite set. We shall denote the number of elements n of A by the symbol $|A|$.

Sets of real and complex numbers are denoted by symbols **R** and **C**, respectively.

2.1.1 Probability Space

The modern axiomatics of probability theory was invented by Andrei Nikolaevich Kolmogorov (one of greatest mathematicians of the 20th century) in 1933, [219], see also [121, 279, 161] and it was a natural finalization of a few hundred years long development of the *set-theoretic model* for probability. A crucial point is representation of events by subsets of some basic set Ω. The collection of subsets representing events should be sufficiently rich to be able to perform set-theoretic operations such as intersection, union and difference of sets.[1] Then one assigns weights (real numbers) to these subsets:

$$A \mapsto \mathbf{P}(A) \tag{2.1}$$

for an event A. They are chosen nonnegative and normalized by 1: $\mathbf{P}(\Omega) = 1$. The weight of a set A which is the disjoint union of sets A_1 and A_2 is equal to the sum of weights of these subsets. The latter property is called *finite additivity*. The map given by (2.1) with mentioned properties is measure-theoretic probability. If the

---

[1] However, it should not be unreasonably large. If too extended a system of subsets is selected, then it may represent "events" that cannot be interpreted in a reasonable way.

A. Khrennikov, *Ubiquitous Quantum Structure*,
DOI 10.1007/978-3-642-05101-2_2, © Springer-Verlag Berlin Heidelberg 2010

basic set Ω is finite[2] one can proceed with this simple definition. However, if the Ω is countable, i.e., it is infinite and its points can be enumerated or "continuous" – e.g., a segment of the real line $\mathbf{R}$ – then finite additivity is not sufficient for creating a fruitful mathematical model. It is extended to so-called σ-additivity (countable additivity). A rich mathematical model is created. However, by proceeding with σ-additivity one should not forget Kolmogorov's remark [219] that σ-*additivity is a purely mathematical and totally nonphysical notion*. It is impossible to perform a real experiment an infinite number of times. In principle, the model based on σ-additivity might produce probabilistic artifacts that have no real interpretation.[3]

We now start rigorous presentation of probability theory. But, in principle, one can read practically the whole book by considering the model based on a finite set Ω, a collection of events represented by all its subsets and finite-additive probability given by assigning weights to points of Ω : $\omega \to \mathbf{P}(\omega)$. For example, the uniform probability is given by equal weights. For $\Omega = \{\omega_1, \ldots, \omega_N\}$, $\mathbf{P}(\omega_j) = 1/N$. Here, for $A \subset \Omega$, $\mathbf{P}(A) = |A|/N$.

Let Ω be a set. We recall that a σ-*algebra* is a system of subsets of Ω that is closed with respect to operations of countable intersection, union and difference of sets and containing Ω and the empty set $\emptyset$.[4]

The simplest example of a σ-algebra is the system consisting of just two sets: Ω and $\emptyset$. However, it is too small to do anything interesting. Another example is given by the family of all subsets of Ω. As was mentioned, such a σ-algebra is useful if the set Ω is finite or even countable. However, if Ω is "continuous", then consideration of all possible subsets as representing events induces visible probabilistic pathologies, see [161] for details. So, the σ-algebra consisting of all subsets (of, e.g., a segment $[a, b]$ of the real line) is too large. One chooses a smaller σ-algebra, the so-called *Borel σ-algebra*. For example, for $\Omega = \mathbf{R}$, it is generated by all half-open intervals: $[\alpha, \beta), \alpha < \beta$. However, in this book we will practically never operate with continuous Ω.

Let Ω be a set and let $\mathcal{F}$ be a σ-algebra of its subsets. A probability measure $\mathbf{P}$ is a map from $\mathcal{F}$ to the segment $[0, 1]$, which is normalized $\mathbf{P}(\Omega) = 1$ and σ-additive

$$\mathbf{P}(A_1 \cup \ldots \cup A_n \cup \ldots) = \mathbf{P}(A_1) + \ldots + \mathbf{P}(A_n) + \ldots,$$

for disjoint sets belonging to $\mathcal{F}$.

By the Kolmogorov axiomatics [219], see also [280], the *probability space* is a triple

$$\mathcal{P} = (\Omega, \mathcal{F}, \mathbf{P}).$$

Points ω of Ω are said to be *elementary events*, elements of $\mathcal{F}$ are *events*, $\mathbf{P}$ is *probability*.

---

[2] Since Ω can contain billions of points, this model is useful in a huge class of applications.

[3] Unfortunately, this point made by Kolmogorov has been totally forgotten.

[4] In some books on probability theory the terminology σ-*field* is used, instead of σ-*algebra*.

Discrete random variables[5] on the Kolmogorov space $\mathcal{P}$ are by definition functions $a : \Omega \to X_a$, where $X_a = \{\alpha_1, \ldots, \alpha_n, \ldots\}$ is a countable set (the *range of values*), such that sets

$$C_\alpha^a = \{\omega \in \Omega : a(\omega) = \alpha\}, \quad \alpha \in X_a, \tag{2.2}$$

belong to $\mathcal{F}$.

It is typically assumed that the range of values X_a is a subset of the real line. We will proceed under this assumption practically everywhere, but sometimes, e.g., in cognitive and psychological modeling, it will be more convenient to consider Boolean labels, e.g. $\alpha = $ yes, no.

We shall use the symbol $RVD(\mathcal{P})$ to denote the space of discrete random variables for the probability space $\mathcal{P}$. The probability distribution of $a \in RVD(\mathcal{P})$ is defined by $\mathbf{P}(a = \alpha) = \mathbf{P}(C_\alpha^a), \alpha \in X_a$, where the set C_α^a is given by (2.2). It is convenient to proceed with a shorter symbol

$$p^a(\alpha) \equiv \mathbf{P}(\omega \in \Omega : a(\omega) = \alpha).$$

We remark that:

$$p^a(\alpha_1) + \ldots + p^a(\alpha_n) + \ldots = 1, \quad p^a(\alpha_n) \geq 0. \tag{2.3}$$

The *average* (mathematical expectation) of a random variable a is defined as

$$\bar{a} \equiv Ea = \alpha_1 \, p^a(\alpha_1) + \ldots + \alpha_n \, p^a(\alpha_n) + \ldots. \tag{2.4}$$

For a family of random variables $a_1, \ldots, a_m$ taking values $\alpha_j^1, \ldots, \alpha_j^m, j = 1, 2, \ldots$, respectively, their joint probability distribution is defined as

$$p^{a_1 \ldots a_m}(\alpha_{j_1}^1, \ldots, \alpha_{j_m}^m) = \mathbf{P}(\omega \in \Omega : a_1(\omega) = \alpha_{j_1}^1, \ldots, a_m(\omega) = \alpha_{j_m}^m). \tag{2.5}$$

We remark that the joint probability is symmetric with respect to permutations; e.g., for two random variables a and b, we have

$$p^{ab}(\alpha, \beta) = \mathbf{P}(\omega \in \Omega : a(\omega) = \alpha, b(\omega) = \beta) = p^{ba}(\beta, \alpha). \tag{2.6}$$

It is an important feature of the Kolmogorov model.

For two random variables a and b *covariance* is defined as

$$\mathrm{cov}(a, b) = E(a - \bar{a})(b - \bar{b}) = \sum_{\alpha\beta}(\alpha - \bar{a})(\beta - \bar{b}) \, p^{ab}(\alpha, \beta). \tag{2.7}$$

---

[5] In Chaps. 1–9 we consider only discrete random variables. In Chaps. 10 and 11 random variables having continuous ranges of values will be used.

It is easy to see that

$$\operatorname{cov}(a, b) = Eab - \bar{a}\bar{b}. \tag{2.8}$$

We remark that covariance is symmetric

$$\operatorname{cov}(a, b) = \operatorname{cov}(b, a). \tag{2.9}$$

2.1.2 Conditional Probability

Kolmogorov's probability model is based on a probability space equipped with the operation of conditioning. In this model *conditional probability* is defined by the well-known *Bayes' formula*

$$\mathbf{P}(B|C) = \mathbf{P}(B \cap C)/\mathbf{P}(C), \quad \mathbf{P}(C) > 0. \tag{2.10}$$

By Kolmogorov's interpretation it is the *probability that an event B occurs under the condition that an event C occurred.* We remark that this formula is a definition, it cannot be derived. The use of this definition of conditional probability is one of fundamental constraints induced by the Kolmogorov model.

We remark that $\mathbf{P}_C(B) \equiv \mathbf{P}(B|C)$ is again a probability measure on $\mathcal{F}$. For a set $C \in \mathcal{F}, \mathbf{P}(C) > 0$, and a (discrete) random variable a, the conditional probability distribution is defined as

$$p_C^a(\alpha) \equiv \mathbf{P}(a = \alpha|C), \quad \alpha \in X_a.$$

We naturally have

$$p_C^a(\alpha_1) + \ldots + p_C^a(\alpha_n) + \ldots = 1, \quad p_C^a(\alpha_n) \geq 0. \tag{2.11}$$

The conditional expectation of a random variable a is defined by

$$E(a|C) = \alpha_1 \, p_C^a(\alpha_1) + \ldots + \alpha_n \, p_C^a(\alpha_n) + \ldots. \tag{2.12}$$

For two random variables a and b, consider conditional probabilities

$$p_{\beta|\alpha} \equiv \mathbf{P}(b = \beta|a = \alpha), \quad p_{\alpha|\beta} \equiv \mathbf{P}(a = \alpha|b = \beta).$$

Following tradition, we will call these probabilities *transition probabilities,* although this terminology might be misleading for our further considerations, see Remark 2.1.

These conditional probabilities can also be written in the form

$$p_{\beta|\alpha} = \mathbf{P}(b = \beta|C_\alpha^a), \quad p_{\alpha|\beta} = \mathbf{P}(a = \alpha|C_\beta^b), \tag{2.13}$$

where, e.g., C_α^a is defined by (2.2). It is, of course, assumed that in the first case $p^a(\alpha) > 0$ and in the second case $p^b(\beta) > 0$.

Remark 2.1 The terminology "transition probabilities" may be rather misleading for this book. Typically $p_{\beta|\alpha}$ is considered as the probability of transition from the state α of some system to another state β of the same system. That is why the symbol $p_{\alpha\beta}$ is typically used, instead of our $p_{\beta|\alpha}$. To come to the standard notation, one should change $p_{\beta|\alpha} \to p_{\alpha\beta}$ and vice versa. However, we will not consider states of systems. For us, $p_{\beta|\alpha}$ is probability of obtaining the value $b = \beta$ of the observable b under the condition that the result $a = \alpha$ was observed in the previous measurement of the observable a. Nevertheless, we will also use the standard terminology – transition probabilities.

It is convenient to use the following definition. A random variable a is said to be *nondegenerate* if

$$p^a(\alpha) > 0 \tag{2.14}$$

for any $\alpha \in X_a$. In future considerations we shall use the matrices of conditional probabilities for successive measurements – transition probabilities

$$\mathbf{P}^{b|a} = (p_{\beta|\alpha}), \quad \mathbf{P}^{a|b} = (p_{\alpha|\beta}). \tag{2.15}$$

The first matrix is well defined if a is nondegenerate and the second if b is nondegenerate. We remark that these matrices are always *left stochastic*. A left stochastic matrix is a square matrix whose columns consist of nonnegative real numbers whose sum is 1. For example, for $\mathbf{P}^{b|a}$, we have that

$$\sum_\beta p_{\beta|\alpha} = \sum_\beta \mathbf{P}(b = \beta | a = \alpha) = \sum_\beta \mathbf{P}_{C_\alpha^a}(b = \beta) = 1 \tag{2.16}$$

for any fixed $a = \alpha$. It is a consequence of the fact that, for any set C of strictly positive probability, $\mathbf{P}_C$ is also a probability measure. In (2.16) we select $C = C_\alpha^a$.

Coming back to Remark 2.1, we notice that in standard notation a matrix of "transition probabilities" is not left, but right stochastic, i.e., all rows sum to 1. We point out the following equality connecting the joint probability distribution of two random variables a and b with their transition probabilities:

$$p^{ab}(\alpha, \beta) = p^a(\alpha)p_{\beta|\alpha} = p^b(\beta)p_{\alpha|\beta} = p^{ba}(\beta, \alpha). \tag{2.17}$$

Conditional probabilities are basic in considerations on *independent random variables,* see Sect. 12.1.1.

In our further considerations one special class of matrices of transition probabilities will play a fundamental role. These are so-called doubly stochastic matrices. We recall that in a *doubly stochastic matrix* all entries are nonnegative and all rows and all columns sum to 1. Of course, in general $\mathbf{P}^{b|a}$ is only left stochastic, not doubly stochastic. The following proposition characterizes random variables inducing doubly stochastic matrices.

Proposition 2.1 *Let a and b be nondegenerate random variables. Then the following conditions are equivalent:*

DS-DS *Both matrices $\mathbf{P}^{a|b}$ and $\mathbf{P}^{b|a}$ are doubly stochastic.*

UD *Random variables are uniformly distributed: $p^a(\alpha) = p^b(\beta) = 1/2$.*

SC *Random variables are "symmetrically conditioned" in the sense*

$$p_{\beta|\alpha} = p_{\alpha|\beta}. \tag{2.18}$$

In the Kolmogorovian model one can guarantee double stochasticity for both $b|a$- and $a|b$-conditioning only for uniformly distributed random variables. This is not the case in non-Kolmogorovian models, e.g., for the quantum probabilistic model, see Sect. 2.4. Here equivalence of conditions **DS-DS** and **SC** plays a crucial role. In fact, the latter is coupled to the symmetry of the scalar product.

Consider now a pair of dichotomous random variables $a = \alpha_1, \alpha_2$ and $b = \beta_1, \beta_2$. The matrix of transition probabilities $\mathbf{P}^{b|a}$ has the form

$$\mathbf{P}^{b|a} = \begin{pmatrix} p_{\beta_1|\alpha_1} & p_{\beta_1|\alpha_2} \\ p_{\beta_2|\alpha_1} & p_{\beta_2|\alpha_2} \end{pmatrix} \tag{2.19}$$

It is doubly stochastic iff $p_{1|1} = p_{2|2}$ and $p_{1|2} = p_{2|1}$, i.e.,

$$\mathbf{P}^{b|a} = \begin{pmatrix} p & 1-p \\ 1-p & p \end{pmatrix} \tag{2.20}$$

In particular, it is automatically symmetric. In this case **SC** is equivalent to the condition $\mathbf{P}^{b|a} = \mathbf{P}^{a|b}$.

2.1.3 Formula of Total Probability

In our further considerations an important role will be played by the *formula of total probability* (FTP). It is a theorem of the Kolmogorov model. Let us consider a countable family of disjoint sets A_k belonging to $\mathcal{F}$ such that their union is equal to Ω and $\mathbf{P}(A_k) > 0$, $k = 1, \ldots$. Such a family is called a *partition* of the space Ω.

Theorem 2.1 *Let $\{A_k\}$ be a partition. Then, for every set $B \in \mathcal{F}$, the following formula of total probability holds:*

$$\mathbf{P}(B) = \mathbf{P}(A_1)\mathbf{P}(B|A_1) + \ldots + \mathbf{P}(A_k)\mathbf{P}(B|A_k) + \ldots. \tag{2.21}$$

Proof We have

$$\mathbf{P}(B) = \mathbf{P}(B \cap \cup_{k=1}^{\infty} A_k) = \sum_{k=1}^{\infty} \mathbf{P}(B \cap A_k) = \sum_{k=1}^{\infty} \mathbf{P}(A_k) \frac{\mathbf{P}(B \cap A_k)}{\mathbf{P}(A_k)}.$$

Especially interesting for us is the case such that a partition is induced by a discrete random variable a taking values $\{\alpha_k\}$. Here

$$A_k = C^a_{\alpha_k} = \{\omega \in \Omega : a(\omega) = \alpha_k\}. \tag{2.22}$$

Let b be another random variable. It takes values $\{\beta_j\}$. For any $\beta \in X_b$, we have

$$\mathbf{P}(b = \beta) = \mathbf{P}(a = \alpha_1)\mathbf{P}(b = \beta | a = \alpha_1) + \ldots + \mathbf{P}(a = \alpha_k)\mathbf{P}(b = \beta | a = \alpha_k) + \ldots \tag{2.23}$$

or in compact notation

$$p^b(\beta) = p^a(\alpha_1) p_{\beta | \alpha_1} + \ldots + p^a(\alpha_k) p_{\beta | \alpha_k} + \ldots . \tag{2.24}$$

2.2 Probabilistic Incompatibility: Bell–Boole Inequalities

If the reader has not yet been excited by *Bell's inequality* and such mysterious consequences of its violation as *quantum nonlocality* and *death of realism* in QM, then I strongly recommend him or her to omit this section as well as Sect. 2.7. Bell's inequality will not play a fundamental role in this book (nevertheless, it will appear in Sect. 9.6).

Bell's inequality is really the central point of modern QM. Therefore the reader may be surprised to find it not in Sect. 2.3, devoted to QM, but in the section devoted to classical probability theory (Kolmogorov's model). However, I think that it is the right place for the appearance of Bell's inequality, i.e., before saying anything about QM. My personal opinion is that this inequality is the standard subject of classical probability theory. Moreover, we will see that Bell-type inequalities appeared in probability theory long before not only Bell's invention, but even the discovery of QM. My main message to the reader is that attempts (which are very popular in modern QM, especially in the quantum information community) to associate Bell-type inequalities with *quantum nonlocality* or *death of realism* are not sufficiently justified. In classical probability theory such inequalities were used for one hundred years (!) without any reference to the mentioned fundamental problems or to QM in general.

2.2.1 Views of Boole, Kolmogorov, and Vorob'ev

In his book [219] Andrei Nikolaevich Kolmogorov emphasized that each experimental arrangement (context) generates its *own probability space*. For him it was totally clear that it is very naive to expect that all experimental contexts can be described by a single (perhaps huge) probability space. In particular, the following problem arises. Suppose that a family of observables, say $\mathcal{O} = \{a_1, a_2, a_3, \ldots\}$, is given. However, it is impossible to measure them all simultaneously. Thus the joint probability distribution is not given. Nevertheless, it is possible to measure some groups of these observables and joint probability distributions for such groups are given.

Is it possible to construct a single probability space serving for the whole family $\mathcal{O}$?

Thus we are interested in the possibility of embedding the family of observables $\mathcal{O}$ into the space of random variables on a single probability space. If the answer is yes, then such observables exhibit *probabilistic compatibility* (PC), and in the opposite case, *probabilistic incompatibility* (PI), see [204] for details.

It seems that G. Boole (the inventor of *Boolean logic* and *Boolean algebra*) was the first to study this problem. He formulated a necessary condition for PC of a family of three dichotomous observables, $a_1, a_2, a_3 = \pm 1$, such that they can be measured pairwise, but not all simultaneously. This condition coincides with the famous Bell's inequality [31], which plays a fundamental role in modern QM![6]. Later the most general problem of PC (i.e., for an arbitrary family of observables) was solved by Soviet mathematician Vorob'ev [302], who applied these results to problems of optimal control and game theory. Unfortunately, Vorob'ev's results were also practically forgotten.[7] Of course, practically complete disregard of the PC problem in the probabilistic community played an extremely negative role in the development of science. In particular, if Bell or at least someone from the quantum community had been aware of the results of Boole or Vorob'ev or at least of Kolmogorov's message, *"context induces a probability space"*, discovery of Bell-type inequalities need not have induced coupling to such mysterious (and nowadays extremely popular) things such as quantum nonlocality or death of realism in QM.

A pragmatic guy [161] could be completely satisfied with recognition that probabilistic data collected in a few incompatible experiments (and this is the case in

---

[6] Boole's results were totally forgotten. Itamar Pitowsky found these results and compared them with Bell's inequality, see [259, 260] and also the preface in [167].

[7] Walter Philipp discovered Vorob'ev's article [302] and together with Karl Hess advertised it a lot [143], in particular during the Växjö series of conferences on quantum foundations, see, e.g., [165, 167, 5, 6]. The main problem of the classical probabilistic community was concentration on mathematical problems related to a single Kolmogorov space, especially various limit theorems. In such a situation even the idea that something could not be embedded in such a space was not especially welcome. Vorob'ev's works were not highly estimated by the Soviet probabilistic community (which was one of the strongest in the world) and, as a result, not by the international community either.

application of Bell's inequality in QM) cannot be described by a single probability space, or in other words, observables are not of the PC-type. Compare the views of Accardi, Aerts, Fine, Fuchs, Garola and Solombrino, Hess and Philipp, Khrennikov, Kupczynski, Larsson, Pitowsky, Rastal, Sozzo, Adenier (see [204] for the corresponding bibliography and this author's book [161] for mathematical details). Then one can try to find sources of PI that are different from quantum nonlocality or death of realism, see [204].

We now discuss the problem in mathematical notation. In principle, it is a repetition of previous considerations, but with mathematical symbols.

Consider a system of three observables $a_i, i = 1, 2, 3$. Suppose for simplicity that they take discrete values and moreover they are dichotomous: $a_i = \pm 1$. Suppose that these observables as well as their pairs can be measured and hence joint probabilities for pairs are well defined: $p^{a_i a_j}(\alpha_i, \alpha_j) \geq 0$ and $\sum_{\alpha_i, \alpha_j = \pm 1} p^{a_i a_j}(\alpha_i, \alpha_j) = 1$.

Question *Is it possible to construct the joint probability distribution, $p^{a_1 a_2 a_3}(\alpha_1, \alpha_2, \alpha_3)$, for any triple of pairwise measurable observables?*

This is the simplest case of the general problem – to find conditions for existence of probability distribution with given *marginal probabilities*. First of all, it is easy to give numerous examples of nonexistence.

Example 2.1 (see [302]) Suppose that

$$\mathbf{P}(a_1 = +1, a_2 = +1) = \mathbf{P}(a_1 = -1, a_2 = -1) = 1/2;$$
$$\mathbf{P}(a_1 = +1, a_3 = +1) = \mathbf{P}(a_1 = -1, a_3 = -1) = 1/2;$$
$$\mathbf{P}(a_2 = +1, a_3 = -1) = \mathbf{P}(a_2 = -1, a_3 = +1) = 1/2.$$

Hence, $\mathbf{P}(a_1 = +1, a_2 = -1) = \mathbf{P}(a_1 = -1, a_2 = +1) = 0$; $\mathbf{P}(a_1 = +1, a_3 = -1) = \mathbf{P}(a_1 = -1, a_3 = +1) = 0$, $\mathbf{P}(a_2 = +1, a_3 = +1) = \mathbf{P}(a_2 = -1, a_3 = -1) = 0$. It is impossible to construct a probability measure which would produce these marginal distributions. We can show this directly [302]. Suppose that one can find a family of real constants $\mathbf{P}(\varepsilon_1, \varepsilon_2, \varepsilon_3)$, $\varepsilon_j = \pm 1$, such that

$$\mathbf{P}(\varepsilon_1, \varepsilon_2, +1) + \mathbf{P}(\varepsilon_1, \varepsilon_2, -1) = \mathbf{P}(a_1 = \varepsilon_1, a_2 = \varepsilon_2), \dots,$$
$$\mathbf{P}(+1, \varepsilon_2, \varepsilon_3) + \mathbf{P}(-1, \varepsilon_2, \varepsilon_3) = \mathbf{P}(a_2 = \varepsilon_2, a_3 = \varepsilon_3).$$

Then it is easy to see that some of these numbers should be negative. In a more fashionable way one can apply Bell's inequality, e.g., for correlations (Sect. 2.2.2) and see that it is violated.

We emphasize that *for mathematicians consideration of Bell-type inequalities did not induce revolutionary reconsideration of the laws of nature. The joint probability distribution does not exist simply because those observables could not be measured simultaneously.*

2.2.2 Bell's and Wigner's Inequalities

Let $\mathcal{P} = (\Omega, \mathcal{F}, P)$ be a Kolmogorov probability space. We recall that *covariance* of two random variables is given by (2.7).

Theorem 2.2 (Bell inequality for covariances) *Let* $a_1, a_2, a_3 = \pm 1$ *be random variables on* $\mathcal{P}$. *Then Bell's inequality*

$$|\langle a_1, a_2 \rangle - \langle a_2, a_3 \rangle| \leq 1 - \langle a_3, a_1 \rangle \qquad (2.25)$$

holds.

The proof of this inequality (in such a rigorous mathematical formulation) can be found, e.g., in [161]; see also the original work of Bell [31] for a proof in the physical framework.

We now turn to Example 2.1. If a_1, a_2, a_3 can be realized on the same probability space, then (2.25) would hold. On the other hand, we have

$$\langle a_1, a_2 \rangle = 1; \quad \langle a_1, a_3 \rangle = 1; \langle a_2, a_3 \rangle = -1.$$

Bell's inequality should imply: $1 - (-1) = 2 \leq 1 - 1 = 0$. We remark that in accordance with Boole we consider Bell's inequality just as a necessary condition for probabilistic compatibility.

We also recall the following simple mathematical result, see Wigner [304]:

Theorem 2.3 (Wigner inequality) *Let* $a_1, a_2, a_3 = \pm 1$ *be arbitrary random variables on a Kolmogorov space* $\mathcal{P}$. *Then the following inequality holds:*

$$\mathbf{P}(a_1 = +1, a_2 = +1) + \mathbf{P}(a_2 = -1, a_3 = +1) \qquad (2.26)$$

$$\geq \mathbf{P}(a_1 = +1, a_3 = +1).$$

Its proof is very simple, see Sect. 12.2. The crucial point is the use of a single probability measure $\mathbf{P}$.

2.2.3 Bell-type Inequalities for Conditional Probabilities

The original *Boole–Bell inequality* served to solve the problem of PC. In its simplest version this problem is based on the assumption that pairwise probability distributions are well defined – observables can be measured pairwise. However, even such an assumption is not always satisfied. Sometimes even for pairs of observables joint measurements are impossible, but it is possible to perform conditional measurements. For example, first the observable a_1 is measured and the result $a_1 = \alpha_1$ is obtained. Then under this condition the observable a_2 is measured. Conditional probability $\mathbf{P}(a_2 = \alpha_2 | a_1 = \alpha_1)$ can be found. The simplest test of PC – the possibility of realizing three observables a_1, a_2, a_3 on a single Kolmogorov probability space – is based on conditional probabilities. As the author of this book noticed, by using Bayes' formula (2.10), i.e., assuming the validity of the Kolmogorovian definition of conditional probability, Wigner's inequality can be easily rewritten as an inequality for conditional probabilities:

Theorem 2.4 (Wigner–Khrennikov inequality) *Let* $a_1, a_2, a_3 = \pm 1$ *be arbitrary random variables on a Kolmogorov space* $\mathcal{P}$. *Then the following inequality holds:*

$$\mathbf{P}(a_1 = +1)\mathbf{P}(a_2 = +1|a_1 = +1) + \mathbf{P}(a_2 = -1)\mathbf{P}(a_3 = +1|a_2 = -1) \quad (2.27)$$

$$\geq \mathbf{P}(a_3 = +1)\mathbf{P}(a_1 = +1|a_3 = +1).$$

Thus if conditional probabilities for a triple of dichotomous observables violate this inequality, they exhibit PI; see Sect. 9.6 for application to game theory.

2.3 Quantum Probabilistic Model

The mathematical formalism of quantum mechanics is the theory of self-adjoint operators on complex Hilbert spaces. The symbols $\mathcal{H}$ and $\langle \cdot, \cdot \rangle$ denote separable complex Hilbert space and the scalar product on it; $\|\psi\| = \sqrt{\langle \psi, \psi \rangle}$ the norm of $\psi \in \mathcal{H}$;

$$S = \{\psi \in \mathcal{H} : \|\psi\| = 1\}$$

is the unit sphere in $\mathcal{H}$. We also consider the set of equivalence classes in the unit sphere S with respect to the equivalence relation: $\psi_1 \sim \psi_2$ iff $\psi_1 = c\psi_2$, where $c \in \mathbf{C}$ and $|c| = 1$. Denote this set by the symbol $\tilde{S}$.

Although real quantum physics is described by infinite-dimensional Hilbert space (of square integrable complex valued functions), quantum information is totally fine with finite dimensional spaces:

$$\mathcal{H}_n = \mathbf{C}^n = \mathbf{C} \times \ldots \times \mathbf{C}. \quad (2.28)$$

Since our considerations relate merely to informational features of the quantum model, we can proceed (practically everywhere) in the same way as in quantum information. The space $\mathcal{H}_n$ is endowed with the scalar product

$$\langle \psi, \phi \rangle = \sum_{j=1}^{n} \psi_j \overline{\phi_j}, \quad \psi = (\psi_1, \ldots, \psi_n), \phi = (\phi_1, \ldots, \phi_n) \in \mathcal{H}_n. \quad (2.29)$$

Self-adjoint operators can be represented by Hermitian matrices, $\hat{a} = (a_{ij})$, such that $a_{km} = \overline{a_{mk}}$, where $z = x + iy \to \bar{z} = x - iy$ is the operation of complex conjugation. The spectrum, $\mathrm{Spec}(\hat{a})$, is nothing else than the set of eigenvalues:

$\widehat{a}\psi = \alpha\psi$. We remark that all eigenvalues are real. Eigenvectors corresponding to the same eigenvalue α form a linear subspace. Its dimension gives the degree of degeneration of α. The orthogonal projector on this subspace is denoted by the symbol P_α^a. It acts similarly to the orthogonal projector to a plane or line in $\mathbf{R}^3$. Of course, the use of complex spaces makes direct geometric illustration impossible even for the space $\mathcal{H}_2$ – it is the four-dimensional real space.

2.3.1 Postulates

The probabilistic model of quantum theory can be formulated as the following series of postulates:

Postulate 1 (The mathematical description of quantum states.) *Quantum (pure) states (wave functions) are represented by normalized vectors ψ (i.e., $\|\psi\|^2 = \langle\psi,\psi\rangle = 1$) of a complex Hilbert space $\mathcal{H}$. Every normalized vector $\psi \in \mathcal{H}$ may represent a quantum state. If a vector ψ corresponding to a state is multiplied by any complex number $c, |c| = 1$, the resulting vector will correspond to the same state.*[8]

The physical meaning of "a quantum state" is not defined by this postulate. It must be provided by a separate postulate; see Postulates 6, 6a.

Postulate 2 (The mathematical description of physical observables.) *A physical observable a is represented by a self-adjoint operator $\widehat{a}$ in complex Hilbert space $\mathcal{H}$. Different observables are represented by different operators.*

Postulate 3 (Spectral) *For a physical observable a which is represented by the self-adjoint operator $\widehat{a}$ we can predict (together with some probabilities) values $\lambda \in$ Spec$(\widehat{a})$ (the spectrum of $\widehat{a}$).*

We restrict our considerations to the simplest self-adjoint operators, which are analogous to discrete random variables. We recall that a self-adjoint operator $\widehat{a}$ has *a purely discrete spectrum* if it can be represented as

$$\widehat{a} = \alpha_1 P_{\alpha_1}^a + \ldots + \alpha_m P_{\alpha_m}^a + \ldots, \quad \alpha_m \in \mathbf{R}, \tag{2.30}$$

where $P_{\alpha_m}^a$ are orthogonal projection operators related to the orthonormal eigenvectors $\{e_{km}^a\}_k$ of $\widehat{a}$ corresponding to the eigenvalues α_m by

$$P_{\alpha_m}^a \psi = \sum_k \langle\psi, e_{km}^a\rangle e_{km}^a, \quad \psi \in \mathcal{H}. \tag{2.31}$$

Here k labels the eigenvectors e_{km}^a which belong to the same eigenvalue α_m of $\widehat{a}$. Thus Spec$(\widehat{a}) = \{\alpha_1, \ldots, \alpha_m, \ldots\}$.

---

[8] Thus states are given by elements of $\tilde{S}$.

Postulate 4 (Born's rule – in formalization of Dirac and von Neumann) *Let a phys-*
ical observable a be represented by a self-adjoint operator $\hat{a}$ *with purely discrete*
spectrum. The probability $\mathbf{P}_\psi(a = \alpha_m)$ *of obtaining the eigenvalue* α_m *of* $\hat{a}$ *for*
measurement of a in a state ψ *is given by*

$$\mathbf{P}_\psi(a = \alpha_m) = \| P_m^a \psi \|^2. \tag{2.32}$$

If the operator $\hat{a}$ has *nondegenerate* (purely discrete) spectrum, then each α_m is
associated with a one-dimensional subspace. The latter can be fixed by selecting any
normalized vector, say e_m^a. In this case orthogonal projectors act simply as

$$P_{\alpha_m}^a \psi = \langle \psi, e_m^a \rangle e_m^a. \tag{2.33}$$

The formula (2.32) takes a very simple form

$$\mathbf{P}_\psi(a = \alpha_m) = |\langle \psi, e_m^a \rangle|^2. \tag{2.34}$$

This is Born's rule in the Hilbert space formalism.

To obtain original Born's rule, one should choose $\mathcal{H}$ as the L_2-space of square
integrable functions, $\psi : \mathbf{R} \mapsto \mathbf{C}$. (We consider a one-dimensional particle.) The
position observable x is represented by the multiplication operator $\hat{x}$

$$\hat{x}(\psi)(x) = x\psi(x). \tag{2.35}$$

This operator has a continuous spectrum. It coincides with the whole real line. So,
this operator is unbounded. Its eigenvectors do not belong to the L_2-space. They are
given by Dirac's δ-functions, i.e., these are generalized eigenvalues, see Dirac [90]

$$\hat{x}(e_\alpha)(x) = \alpha e_\alpha(x), \quad \alpha \in \mathbf{R}, \tag{2.36}$$

where $e_\alpha(x) = \delta(x - \alpha)$. One can reasonably define paring[9]

$$\langle \psi, e_\alpha \rangle = \psi(\alpha). \tag{2.37}$$

Then the rule (2.34) gives

$$\mathbf{P}_\psi(x = \alpha) = |\psi(\alpha)|^2. \tag{2.38}$$

---

[9] In fact, the situation is little bit more complicated from the mathematical viewpoint. In the
rigorous mathematical framework, elements of the L_2-space are given by equivalent classes of
functions. Two functions belong to the same class if the measure of points where they are distinct
is equal to zero. To proceed rigorously, one should select a subspace in the L_2-space and consider
Dirac's delta function and its shifts $e_\alpha(x) = \delta(x - \alpha)$ as continuous linear functionals on this
subspace. This can be done in the framework of distribution theory. Here paring (2.37) is nothing
else than action of the functional e_α to the test function ψ. However, physicists typically do not
pay attention to such mathematical problems.

*Remark 2.2 (*Origin of Born's rule.*)* This rule was invented in the following way. Originally Schrödinger considered the ψ-function as a classical field – similar to the electromagnetic field. The quantity $E(\alpha) = |\psi(\alpha)|^2$ is the energy density of this field. Born invented the rule (2.38) by criticizing Schrödinger's interpretation. Instead of the energy density, he considered this quantity as the probability density. The latter induces automatically the normalization condition

$$1 = \int_{-\infty}^{+\infty} |\psi(\alpha)|^2 = \langle \psi, \psi \rangle$$

which was absent in Schrödinger's model. After a few years of struggle, Schrödinger gave up and kept to Born's interpretation.

In the same way one can consider momentum measurement. Schrödinger defined the momentum operator as

$$\widehat{p}(\psi)(x) = -i\frac{d}{dx}\psi(x). \tag{2.39}$$

(We eliminate the Planck constant from consideration by choosing the appropriate system of units.) It is easy to see that its spectrum is also continuous and it coincides with $\mathbf{R}$. Its generalized eigenfunctions can be easily found from the equation

$$-i\frac{d}{dx}e_\beta^p(x) = \beta e_\beta^p(x), \quad \beta \in \mathbf{R}.$$

Thus $e_\beta^p(x) = e^{i\beta x}$. Thus by (2.34)

$$\mathbf{P}_\psi(p = \beta) = |\langle \psi, e_\beta^b \rangle|^2. \tag{2.40}$$

By taking into account that

$$\langle \psi, e_\beta^b \rangle = \int_{-\infty}^{+\infty} \psi(x)e^{-i\beta x}dx = \tilde{\psi}(\beta)$$

is the Fourier transform of ψ, we write Born's rule for the momentum measurement as

$$\mathbf{P}_\psi(p = \beta) = |\tilde{\psi}(\beta)|^2, \tag{2.41}$$

cf. (2.38).

*Remark 2.3 (*Classical description of quantum measurements.*)* For any state ψ, each quantum observable $\widehat{a}$ can be represented as a classical random variable. In the discrete case we take $\Omega = \{\alpha_1, \ldots, \alpha_m, \ldots\} \equiv \mathrm{Spec}(\widehat{a})$, the σ-algebra consists of all subsets of Ω, and the probability measure is defined as $\mathbf{P}(A) = \sum_{\alpha_m \in A} \mathbf{P}_\psi$

$(a = \alpha_m)$, where $\mathbf{P}_\psi(a = \alpha_m)$ is given by Born's rule. Thus each concrete quantum measurement can be described classically. Problems arise only when one tries to describe classically data collected for a few incompatible observables. We remark that such attempts contradict Kolmogorov's ideology [219]. Kolmogorov emphasized that each probability space is determined by the corresponding complex of experimental conditions (context). The same message came from Bohr, who pointed out that the whole experimental arrangement should be taken into account and whose *principle of complementarity* supports Kolmogorovian ideology. For example, the impossibility of embedding the collection of probabilities for the position and momentum measurements (for all possible quantum states) into a single probability space is often considered as a new astonishing probabilistic situation. However, Kolmogorov's ideology implies that attempts at such an embedding have no justification – since the position and momentum measurements for a quantum system cannot be performed in a single experimental setting.

By using Born's rule (2.32) and the classical probabilistic definition of average (2.4), it is easy to see that the average value of an observable a in a state ψ belonging to the domain of definition of the operator $\hat{a}$ is given by

$$\langle a \rangle_\psi = \langle \hat{a}\,\psi, \psi \rangle. \tag{2.42}$$

Postulate 5 (Time evolution of wave function.) *Let $\hat{H}$ be the Hamiltonian of a quantum system, i.e., the self-adjoint operator corresponding to the energy observable. The time evolution of the wave function $\psi \in \mathcal{H}$ is described by the Schrödinger equation*

$$i\frac{d}{dt}\psi(t) = \hat{H}\psi(t) \tag{2.43}$$

with the initial condition $\psi(0) = \psi$.

2.3.2 Quantization

We remark that the operators of position and momentum, $\hat{x}$ and $\hat{p}$, see (2.35) and (2.39), do not commute and they satisfy Heisenberg's *canonical commutation relation*

$$[\hat{x}, \hat{p}] = i. \tag{2.44}$$

Consider any real-valued function on the classical phase space, i.e., a function of classical coordinate and momentum, $f(x, p)$. The quantization procedure is the map

$$f \mapsto \hat{f} = f(\hat{x}, \hat{p}). \tag{2.45}$$

In general, it is a tricky mathematical problem to define a function of two noncommuting operators. It is typically done by using the calculus of pseudo-differential operators.[10]

However, in the simplest case the operator of energy $\widehat{H}$ can be easily defined. Consider a classical particle with the mass m moving in the potential $V(x)$. Its Hamiltonian function (representing classical energy of this particle) is given by

$$H(x, p) = \frac{p^2}{2m} + V(x). \tag{2.46}$$

Quantization gives us the operator

$$\widehat{H} = H(\widehat{x}, \widehat{p}) = \frac{\widehat{p}^2}{2m} + V(x). \tag{2.47}$$

2.3.3 Interpretations of Wave Function

Now we are going to discuss one of the most important and complicated notions of quantum mechanics: the notion of a quantum state. There are two main points of view, which are formulated in the following postulates.

Postulate 6 (The ensemble interpretation.) *A wave function provides a description of certain statistical properties of an ensemble of similarly prepared quantum systems.*

This interpretation is upheld, for example by Einstein, Popper, Blokhintsev, Margenau, Ballentine, Klyshko, and in recent years by, e.g., de Muynck, De Baere, Holevo, Santos, Khrennikov, Nieuwenhuizen, Adenier and many others.

Postulate 6a (The Copenhagen interpretation.) *A wave function provides a complete description of an individual quantum system.*

This interpretation was supported by a great variety of scientists, from Schrödinger, in his original attempt to identify the electron with a wave function solution of his equation, to the proponents of the several versions of the Copenhagen interpretation (for example, Heisenberg, Bohr, Pauli, Dirac, von Neumann, Landau, Fock and, in recent years, e.g., Greenberger, Mermin, Lahti, Peres, Summhammer[11]). Nowadays

---

[10] See [160] for the most general presentation of quantization procedure on the mathematical level of rigorousness, including both bosons and fermions as well as supersymmetric systems, quantum field theory, strings and superstrings and corresponding string field theories; see [158, 159] for operator quantization over non-Archimedean (in particular, p-adic) number fields.

[11] There is an interesting story about the correspondence between Bohr and Fock on the individual interpretation. This story was told to me by a former student of Fock, who pointed out that one of the strongest supporters of this interpretation was Vladimir A. Fock, and that even though Bohr himself had doubts about its consistency, Fock demonstrated to Bohr inconsistency in the Einsteinian ensemble interpretation. Thus interpretation, which is commonly known as the *Copenhagen interpretation*, might just as well be called the "*Leningrad interpretation.*"

the individual interpretation is extremely popular, especially in quantum information and computing.

Instead of Einstein's terminology "*ensemble interpretation*", Ballentine [25, 26] used the terminology "*statistical interpretation.*" However, Ballentine's terminology is rather misleading, because the term "statistical interpretation" was also used by von Neumann for individual randomness! For him "statistical interpretation" had a meaning that is totally different from Ballentine's "ensemble-statistical interpretation." John von Neumann wanted to emphasize the difference between deterministic (Newtonian) classical mechanics, in which the state of a system is determined by values of two observables (position and momentum), and quantum mechanics, in which the state is determined not by values of observables, but by probabilities. We shall follow Albert Einstein and use the terminology "*ensemble interpretation.*"

Finally, we point out recent papers concerning the foundations and, in particular, various interpretations of quantum mechanics: [3–6, 13, 21, 106, 116, 130, 4, 17, 21, 27, 45, 47, 66, 77, 79, 83–85, 106, 116, 117, 130, 119, 127, 142–144, 148, 161– 214, 215, 225, 226, 261– 264, 272, 284–292].

2.4 Quantum Conditional Probability

As in the classical Kolmogorov probabilistic model, Born's postulate should be completed by a definition of conditional probability. We present the contemporary definition that is conventional in quantum logic [32] and quantum information theory.

Definition 2.1 Let physical observables a and b be represented by self-adjoint operators with purely discrete (possibly degenerate) spectra:

$$\widehat{a} = \sum_m \alpha_m P^a_{\alpha_m}, \ \widehat{b} = \sum_m \beta_m P^b_{\beta_m} \tag{2.48}$$

Let ψ be a pure state and let $P^a_{\alpha_k} \psi \neq 0$. Then the probability of obtaining the value $b = \beta_m$ under the condition that the value $a = \alpha_k$ was observed in the preceding measurement of the observable a on the state ψ is given by

$$\mathbf{P}_\psi(b = \beta_m | a = \alpha_k) \equiv \frac{\| P^b_{\beta_m} \ P^a_{\alpha_k} \psi \|^2}{\| P^a_{\alpha_k} \ \psi \|^2} \tag{2.49}$$

Let the operator $\widehat{a}$ have a nondegenerate spectrum, i.e., for any eigenvalue α the corresponding eigenspace (i.e., generated by eigenvectors with $\widehat{a}\psi = \alpha\psi$) is one dimensional. We can write

$$\mathbf{P}_\psi(b = \beta_m | a = \alpha_k) = \| P^b_{\beta_m} e^a_k \|^2 \tag{2.50}$$

(here $\widehat{a}e_k^a = \alpha_k e_k^a$). Thus the conditional probability in this case does not depend on the original state ψ. We can say that the memory of the original state has been destroyed. If also the operator $\widehat{b}$ has a nondegenerate spectrum then we have $\mathbf{P}_\psi(b = \beta_m | a = \alpha_k) = |\langle e_m^b, e_k^a\rangle|^2$ and $\mathbf{P}_\psi(a = \alpha_k | b = \beta_m) = |\langle e_k^a, e_m^b\rangle|^2$. By using symmetry of the scalar product we obtain:

Proposition 2.2 *Let both operators $\widehat{a}$ and $\widehat{b}$ have purely discrete nondegenerate spectra and let $P_k^a \psi \neq 0$ and $P_m^b \psi \neq 0$. Then conditional probability is symmetric and it does not depend on the original state ψ :*

$$\mathbf{P}_\psi(b = \beta_m | a = \alpha_k) = \mathbf{P}_\psi(a = \alpha_k | b = \beta_m) = |\langle e_m^b, e_k^a\rangle|^2.$$

We remark that classical (Kolmogorov–Bayes) conditional probability is not symmetric, except in very special situations; the same is valid for my general contextual probabilistic model, see Chapt. 3. Thus *QM is described by a very specific probabilistic model.*

Consider two nondegenerate observables. Set $p_{\beta|\alpha} = \mathbf{P}(b = \beta | a = \alpha)$. The matrix of transition probabilities $\mathbf{P}^{b|a}$, see (2.15) for the definition (but do not forget that transition probabilities are no longer defined by Bayes' rule!), is not only *stochastic* but *doubly stochastic.* It is easy to see that

$$\sum_\alpha p_{\beta|\alpha} = \sum_\alpha |\langle e_\beta^b, e_\alpha^a\rangle|^2 = \langle e_\beta^b, e_\beta^b\rangle = 1.$$

Double stochasticity is also a very specific property of quantum probability, cf. the Kolmogorovian model and my model Chap. 3. In fact, condition **DS-DS** holds: both matrices of transition probabilities $\mathbf{P}^{a|b}$ and $\mathbf{P}^{b|a}$ are doubly stochastic. Moreover, any pair of quantum observables (with nondegenerate spectra) satisfies to condition **SC**; they are "symmetrically conditioned", see (2.18).

In the quantum framework *independent observables* are considered in Sect. 12.1.2.

2.5 Interference of Probabilities in Quantum Mechanics

We will show that quantum probabilistic calculus violates the conventional FTP, see Sect. 2.1.3.

Let $\mathcal{H}_2 = \mathbf{C} \times \mathbf{C}$ be the two-dimensional complex Hilbert space and let $\psi \in \mathcal{H}_2$ be a quantum state. Let us consider two dichotomous observables $b = \beta_1, \beta_2$ and $a = \alpha_1, \alpha_2$ represented by self-adjoint operators $\widehat{b}$ and $\widehat{a}$, respectively (one may consider simply Hermitian matrices). Let $e^b = \{e_\beta^b\}$ and $e^a = \{e_\alpha^a\}$ be two orthonormal bases consisting of eigenvectors of the operators. The state ψ can be represented in the two ways

$$\psi = c_1 e_1^a + c_2 e_2^a, \quad c_\alpha = \langle \psi, e_\alpha^a\rangle; \tag{2.51}$$

$$\psi = d_1 e_1^b + d_2 e_2^b, \quad d_\beta = \langle \psi, e_\beta^b \rangle. \tag{2.52}$$

By Postulate 4 we have

$$\mathbf{P}(a = \alpha) \equiv \mathbf{P}_\psi(a = \alpha) = |c_\alpha|^2; \tag{2.53}$$

$$\mathbf{P}(b = \beta) \equiv \mathbf{P}_\psi(b = \beta) = |d_\beta|^2. \tag{2.54}$$

The possibility of expanding one basis with respect to another basis induces connection between the probabilities $\mathbf{P}(a = \alpha)$ and $\mathbf{P}(b = \beta)$. Let us expand the vectors e_α^a with respect to the basis e^b

$$e_1^a = u_{11} e_1^b + u_{12} e_2^b; \tag{2.55}$$

$$e_2^a = u_{21} e_1^b + u_{22} e_2^b, \tag{2.56}$$

where $u_{\alpha\beta} = \langle e_\alpha^a, e_\beta^b \rangle$. Thus $d_1 = c_1 u_{11} + c_2 u_{21}$, $d_2 = c_1 u_{12} + c_1 u_{22}$. We obtain the *quantum rule* for transformation of probabilities

$$\mathbf{P}(b = \beta) = |c_1 u_{1\beta} + c_2 u_{2\beta}|^2. \tag{2.57}$$

On the other hand, by the definition of quantum conditional probability, see (2.49), we obtain

$$\mathbf{P}(b = \beta | a = \alpha) \equiv \mathbf{P}_\psi(b = \beta | a = \alpha) = |\langle e_\alpha^a, e_\beta^b \rangle|^2. \tag{2.58}$$

By combining (2.53), (2.54) and (2.57), (2.58) we obtain the *quantum formula of total probability – the formula of interference of probabilities*:

$$\mathbf{P}(b = \beta) = \sum_\alpha \mathbf{P}(a = \alpha) \mathbf{P}(b = \beta | A = \alpha)$$

$$+ 2 \cos \theta \sqrt{\mathbf{P}(a = \alpha_1) \mathbf{P}(b = \beta | a = \alpha_1) \mathbf{P}(a = \alpha_2) \mathbf{P}(b = \beta | a = \alpha_2)} \tag{2.59}$$

In general $\cos \theta \neq 0$. Thus the quantum FTP does not coincide with the classical FTP (2.23) which is based on Bayes' formula

$$\mathbf{P}(b = \beta) = \sum_\alpha \mathbf{P}(a = \alpha) \mathbf{P}(b = \beta | a = \alpha). \tag{2.60}$$

2.6 Contextual Point of View of Interference

The difference between the quantum rule (2.59) and the classical rule (2.60) is not surprising. As was pointed out in Remark 2.2, there are no reasons to expect that data obtained for observables a and b which could not be jointly measured can be described by a single Kolmogorov probability space. However, the classical FTP, see Sect. 2.1.3, was derived under the assumption that both observables can be represented by random variables belong to the same Kolmogorov space.[12]

The crucial point is that one cannot use the same symbol $\mathbf{P}$ to denote all probabilities in (2.59). In one formula, (2.59), one combines probabilistic data obtained in four different experiments (experimental contexts):

a) measurement of the observable a under the complex of physical conditions (context) C which is represented by the initial state ψ;
b) measurement of the observable b under the same context C;

After performing the a-measurement one can create through selection procedures C_{α_1} and C_{α_2} (selections of systems with respect to the values $a = \alpha_1$ and $a = \alpha_2$) two new ensembles of systems S_{α_1} and S_{α_2}. In quantum mechanics (with the ensemble interpretation) these ensembles are represented by the eigenvectors e_1^a, e_2^a of the operator $\hat{a}$. Therefore we can perform the b-measurement for two new contexts:

a1) measurement of the observable b under the complex of physical conditions (context) C_{α_1} which is represented by the state e_1^a;
a2) measurement of the observable b under the complex of physical conditions (context) C_{α_2} which is represented by the state e_2^a.

The a)-experiment gives probabilities $\mathbf{P}_\psi(a = \alpha)$; the b)-experiment – $\mathbf{P}_\psi(b = \beta)$; the a1)-experiment – $\mathbf{P}_{e_1^a}(b = \beta)$; the a2)-experiment – $\mathbf{P}_{e_2^a}(b = \beta)$.

What could be the reason to assume that we can use a single probability measure $\mathbf{P}$ in all these experiments?

2.7 Bell's Inequality in Quantum Physics

As was pointed out in Sect. 2.2, inequalities of Boole–Bell type provide necessary conditions for probabilistic compatibility (PC) of families of observables and, hence, their violations provide sufficient conditions for probabilistic incompatibility (PI). As was first pointed out by Bell, see [31] for details, quantum formalism

---

[12] We remark that Feynman [105] considered violation of FTP in the two-slit experiment as violation of the laws of classical probability. For him it was an exhibition of special, even mystical, properties of quantum systems. A similar comment by d'Espagnat on violation of FTP can be found in [87].

predicts the existence of such quantum states[13] that inequality (2.25) is violated for a special choice of a family of pairwise measurable observables.[14] Thus these observables are of the PI-type.

In any domain of science, one should look for special roots of PI. In particular, in physics Bell found two possible roots: quantum nonlocality and death of realism. Moreover, he was sure that one can still proceed in QM by using the realistic description in its strongest (Einsteinian) form: assigning values of observables to the state of a quantum system before measurement.[15] In principle, one cannot exclude that he found the right possible roots.

My approach is essentially more general. By considering the problem from the PI viewpoint, we can look for other roots of PI, which need not coincide with those proposed by Bell. One can still keep to realism and locality. PI can arise from, e.g., taking into account parameters of measurement devices (so considering values of observables as depending on internal states not only of systems, but also of measurement devices[16]), or from unfair sampling; details can be found in [204, 7].

Moreover, Bell-type inequalities for probability distributions (or covariances) of pairwise measurements are not the simplest tests of PC. As was mentioned in Sect. 2.2.3, PC can be tested by conditional measurements of three observables by using the Wigner–Khrennikov inequality (2.27). It is easy to see [214] that this inequality is violated for specially selected projections of spin or polarization. Conditional measurements, e.g., spin projections to one direction and then to another, can be performed on a single particle. Unlike Bell's original scheme, we need not consider pairs of entangled particles. Hence, PI of spin or polarization projections take place even for a single particle. It is completely clear that the source of PI is the impossibility of measuring these observables simultaneously. It would be surprising if PI for spin or polarization projections derived by using Bell's original inequality for entangled pairs has another explanation, e.g., nonlocality. By operating with the Wigner–Khrennikov inequality for conditional probabilities one can see how artificial Bell's appeal to nonlocality was.

---

[13] These are so called EPR-type states, see Einstein, Podolsky, Rosen [99] for details.

[14] For example, spin or polarization projections to specially chosen directions.

[15] Here by state we understood "prequantum state", hidden variable, λ. Thus, first of all, J. Bell was sure that QM does not provide the complete description of phenomena. As well as Einstein, he was sure that one can finally find a better description of physical reality than given by QM. The reason of Bell's belief in naive Einsteinian realism were precise correlations (anti-correlations) exhibited by measurements for EPR-type states. Thus Bell was sure that violation of inequality (2.25) implies nonlocality. For him, the best model of prequantum reality was given by Bohmian mechanics. Later, as is often happen in science, majority of people combine nonlocality with rejection of realism. The monster of mysterious "quantum nonlocality" was born. It is clear that Bell would not be happy with such an interpretation of his studies. However, it is clear as well that Einstein would not be happy with nonlocal realism. His reaction to creation of Bohmian mechanics was negative.

[16] Such sort of realism differs from naive *Einsteinian realism* and it is closer to Bohr's views; cf. also with Accardi's chameleon effect [1, 4] and Ohya's adaptive dynamics [245, 246].

2.8 Växjö Interpretation of Quantum Mechanics

The Växjö interpretation [177] is a variant of the ensemble interpretation, Postulate 6:

A wave function provides a description of certain statistical properties of an ensemble of similarly prepared quantum systems.

However, "properties" are not Einsteinian properties, which can be assigned to a system before measurement. Properties should be understood in Bohr's sense: as results of interaction of systems with measurement devices. However, unlike Bohr, I do not claim that QM is complete and it is in principle impossible to provide a finer description of reality, e.g., by taking into account internal states of measurement devices, see [214, 184, 191].

Chapter 3
Contextual Probabilistic Model – Växjö Model

A contextual probabilistic model – providing a general description of probabilistic data, classical as well as quantum – will be presented. Moreover, the classical Kolmogorovian model [219] and the quantum Dirac–von Neumann model [90, 301], see Sect. 2.3, are only very special cases of our contextual model. The latter describes probabilistic data that cannot be described by either classical or conventional quantum models.

3.1 Contextual Description of Observations

A fundamental notion of my model is *context*. It is a complex of (e.g. physical or biological) conditions. Construction of the model starts with selection of a family of contexts $\mathcal{C}$. The next step is selection of a family of *observables* $\mathcal{O}$. Any observable $a \in \mathcal{O}$ can be measured under context $C \in \mathcal{C}$.

Denote observables by Latin letters, $a, b, ...$, and their values by Greek letters, $\alpha, \beta, ...$ For an observable $a \in \mathcal{O}$, denote the set of its possible values ("spectrum") by the symbol X_a. To simplify considerations, we will consider only discrete observables. We remark that our general model does not contain systems, e.g., physical ones.

3.1.1 Contextual Probability Space and Model

Definition 3.1 *A contextual probability space is a triple $\mathcal{P}_{cont} = (\mathcal{C}, \mathcal{O}, \pi)$, where elements of $\mathcal{C}$ and $\mathcal{O}$ are interpreted as contexts and observables, and elements of π are the corresponding probability distributions[1].*

Here $\pi = \{ p_C^a \}$, $C \in \mathcal{C}, a \in \mathcal{O}$. For any $\alpha \in X_a$,

$$p_C^a(\alpha) \equiv \mathbf{P}(a = \alpha | C) \tag{3.1}$$

---

[1] In our case these are simply discrete probability measures.

A. Khrennikov, *Ubiquitous Quantum Structure*,
DOI 10.1007/978-3-642-05101-2_3, © Springer-Verlag Berlin Heidelberg 2010

is the probability of obtaining the value $a = \alpha$ for observation of a under context C. We have

$$p_C^a(\alpha_1) + ... + p_C^a(\alpha_n) + ... = 1, \qquad p_C^a(\alpha_n) \geq 0. \tag{3.2}$$

We prefer to call probabilities (3.1) *contextual probabilities*.[2] For any context $C \in \mathcal{C}$, we consider the set of probabilities

$$W(\mathcal{O}, C) = \{\mathbf{P}(a = \alpha | C) : a \in \mathcal{O}, \quad \alpha \in X_a\}. \tag{3.3}$$

Definition 3.2 Contextual expectation $E[a|C]$ of an observable $a \in \mathcal{O}$ with respect to context $C \in \mathcal{C}$ is given by

$$\bar{a}_C = E[a|C] = \alpha_1 \, p_C^a(\alpha_1) + ... + \alpha_n \, p_C^a(\alpha_n) + ... \tag{3.4}$$

I now compare (3.2) and (3.4) with the Kolmogorovian model, namely, with (2.3) and (2.4). In my model the probability distribution of an observable a depends on a context (of observations) C. In Kolmogorov's approach equations (2.3) and (2.4) are based on the "absolute probability distribution" $\mathbf{P}$. However, the Kolmogorovian model also provides a possibility (although rather restricted) of inventing contextual dependence of probabilities, namely, through contextual interpretation of conditional probability. Formally, (3.2) and (3.4) coincide with (2.11), (2.12). The main difference is that in the Kolmogorovian model it is assumed that all probabilities can be produced from a single probability $\mathbf{P}$ with the aid of Bayes' formula.

Although the definition of a contextual probability space does not involve systems, in the majority of applications one can assume that measurements are performed on *systems*, physical, biological, social. In such a case, it is useful to assume that each context C can be represented by an ensemble of systems S_C. These are systems that have interacted with C. They can be considered as representing features of C. Typically, a system represents only a "part of the features" of C. To represent context C, a sufficiently large (in principle, infinitely large) ensemble S_C should be used, see [214] for formalization.

To create a fruitful model, we postulate the existence of contexts inducing transition probabilities $\mathbf{P}(b = \beta | a = \alpha)$ for pairs of observables.

Definition 3.3 (Växjö model) A contextual probability model is a contextual probability space $\mathcal{P}_{\text{cont}} = (\mathcal{C}, \mathcal{O}, \pi)$ such that $\mathcal{C}$ contains a special subfamily of contexts $\{C_\alpha^a\}_{a \in \mathcal{O}, \alpha \in X_a}$ which are interpreted as $[a = \alpha]$-selection contexts. Context C_α^a corresponds to the selection with respect to the result $a = \alpha$. Moreover, it is assumed that C_α^a satisfies the condition

---

[2] It is possible to call them conditional probabilities as in Kolmogorov's model. Unlike the latter, contextual probability (3.1) is not the probability that an event, say B, occurs under the condition that another event, say C, occurred. Contextual probability is the probability of obtaining the result $a = \alpha$ under context C.

$$\mathbf{P}(a = \alpha | C_\alpha^a) = 1. \tag{3.5}$$

It is assumed that, for each observable $a \in \mathcal{O}$ and its value α, the selection context C_α^a is uniquely determined – in the class of contexts $\mathcal{C}$ of the model. To simplify notation, we shall often use the symbol C_α instead of C_α^a (when such a notation is not ambiguous). Both Kolmogorov's classical model and the Dirac–von Neumann quantum model can be represented as contextual models, see Sect. 12.4.

3.1.2 Selection Contexts; Analogy with Projection Postulate

The most natural interpretation of selection contexts can be given for the special class of models in which observables are considered as *observables on systems*. In this case the context C_α^a consists of the a-measurement procedure and the post-measurement selection of systems for which the result $a = \alpha$ was obtained. In this book we will be interested merely in cognitive applications. In this framework a is a question which is posed to a group of people. The C_α^a is selection of people who answered $a = \alpha$ (e.g., $\alpha = +1$, i.e. yes, $\alpha = -1$, i.e. no) and creation of a new group. Then people from this new group can be asked another question, say b.

We now comment on condition (3.5). It implies that in a measurement of a under the complex of conditions C_α^a the value $a = \alpha$ is obtained with probability 1. It can be considered as a *contextual probabilistic version of the von Neumann–Lüders projection postulate*, see Sect. 12.3.[3] In cognitive models, systems take a sort of responsibility for their answers to questions belonging to the family $\mathcal{O}$. Thus observables considered in the Växjö model are generalizations of quantum observables.

3.1.3 Transition Probabilities, Reference Observables

Let $a, b \in \mathcal{O}$ and let $\alpha \in X_a, \beta \in X_b$. We consider the $[a = \alpha]$-selection context C_α. The corresponding contextual probabilities

$$p_{\beta|\alpha} \equiv \mathbf{P}(b = \beta | a = \alpha) = \mathbf{P}(b = \beta | C_\alpha)$$

will play an important role in further considerations. They are called *transition probabilities*. As was pointed out, we have simply borrowed the standard terminology. In fact, it would be more natural to call them $b|a$-contextual probabilities. We will use matrices of transition probabilities, for pairs of observables $a, b \in \mathcal{O}$, $\mathbf{P}^{b|a} = (p_{\beta|\alpha})$.

---

[3] By obtaining the fixed result $a = \alpha$ we can be sure that in the process of measurement the system was transformed to such a state that if the a-measurement is performed once again (inside of a sufficiently short interval of time), then the same result $a = \alpha$ will be obtained once again with probability 1, cf. von Neumann [301].

Let $C \in \mathcal{C}$. We complete the probabilistic data $W(\mathcal{O}, C)$, see (3.3), by the data contained in the matrices of transition probabilities $\mathbf{P}^{b|a}$ for all pairs $a, b \in \mathcal{O}$. We obtain a collection of contextual probabilities that will be denoted by the symbol

$$D(\mathcal{O}, C). \tag{3.6}$$

We shall often take a subset $\mathcal{O}'$ of $\mathcal{O}$ and consider the collection of probabilistic data about contexts given by observables belonging to $\mathcal{O}'$. This collection is denoted by the symbol $D(\mathcal{O}', C)$. Typically $\mathcal{O}'$ will consists of two observables, say a and b. Our aim is to create a quantum-like representation of contexts by using just a pair of observables (which are analogues of position and momentum observables). Such observables are called *reference observables.*

By collecting the probabilistic data $D(\mathcal{O}, C)$ for all contexts $C \in \mathcal{C}$ we obtain the collection of data $\mathcal{D}(\mathcal{O}, \mathcal{C}) = \cup_{C \in \mathcal{C}} D(\mathcal{O}, C)$, which completely characterizes the contextual probabilistic model. Thus any model can be symbolically written as $M = (\mathcal{C}, \mathcal{O}, \mathcal{D}(\mathcal{O}, \mathcal{C}))$.

The main considerations will be based on pairs of dichotomous reference observables: $\mathcal{O}' = \{a, b\}$ and $a = \alpha_1, \alpha_2, \ b = \beta_1, \beta_2$. Here

$$D(a, b, C) = \{p_C^a(\alpha), p_C^b(\beta), p_{\alpha|\beta}, p_{\beta|\alpha}\}.$$

In further considerations we will be interested in pairs of reference observables such that both matrices of transition probabilities $\mathbf{P}^{b|a}$ and $\mathbf{P}^{a|b}$ (or $b|a$ and $a|b$ contextual probabilities) are doubly stochastic. As in the Kolmogorovian case, we call two observables *symmetrically conditioned* if condition **SC** (2.18) holds. This condition implies double stochasticity. However, unlike the Kolmogorovian case, double stochasticity even of both matrices of transition probabilities $\mathbf{P}^{b|a}$ and $\mathbf{P}^{a|b}$ does not imply **SC**. We recall that in QM the condition **SC** holds for any pair of observables with nondegenerate spectra. It is a consequence of the symmetry of the scalar product.

3.1.4 Covariance

In the Kolmogorovian model, it is possible to define covariance of two random variables, see (2.7). This definition is based on the joint probability distribution $p^{ab}(\alpha, \beta) = \mathbf{P}(a = \alpha, b = \beta)$. The possibility of joint measurement is assumed. In my contextual probability model such a possibility is not assumed. Thus the joint probability distribution of two observables $a, b \in \mathcal{O}$ need not be defined. Nevertheless, we can proceed by mimicking formula (2.17), a consequence of Bayes' rule in the Kolmogorovian model. We proceed formally by using it as a definition of "joint probability". We set

$$p_C^{ab}(\alpha, \beta) = p_C^a(\alpha) p_{\beta|\alpha}. \tag{3.7}$$

We have $\sum_{\alpha,\beta} p_C^{ab}(\alpha, \beta) = 1$. Thus it is really probability (discrete probability measure on the collection of all subsets of $\Omega = X_a \times X_b$).

Let $a, b \in \mathcal{O}$ be two observables. It is not assumed that they can be jointly measured. Consider any function $h : X_a \times X_b \to \mathbf{R}$. We define the average

$$E[h(a, b)|C] \equiv E_{p_C^{ab}} h(a, b) = \sum_{\alpha,\beta} h(\alpha, \beta) p_C^{ab}(\alpha, \beta). \qquad (3.8)$$

We now define covariance, cf. (2.7). By definition

$$\mathrm{cov}_C(a, b) = E[(a - \bar{a}_C)(b - \bar{b}_C)|C], \qquad (3.9)$$

cf. (2.8), Sect. 2.1.1. In general,

$$\mathrm{cov}_C(b|a) \neq E_{p_C^{ab}}(ab) - \bar{a}_C \bar{b}_C. \qquad (3.10)$$

In the same way we define another "joint probability distribution", which is based on $a|b$-conditioning $p_C^{ba}(\beta, \alpha) = p_C^b(\beta) p_{\alpha|\beta}$ and the corresponding covariance $\mathrm{cov}_C(a|b)$. Unlike the Kolmogorovian model, cf. (2.17) and (2.9), these joint probability distributions and covariances are not equal.

Suppose now that, for some context C,

$$p_C^{ba}(\beta, \alpha) = p_C^{ab}(\alpha, \beta). \qquad (3.11)$$

Then it is possible to construct a Kolmogorov probability space $\mathcal{P}_C$ and realize the obseravbles a and b by random variables: $\Omega = X_a \times X_b$, $\mathcal{F}$ is the collection of its subsets, and probability $\mathbf{P}(A) = \sum_{(\alpha,\beta)\in A} p_C^{ab}(\alpha, \beta)$. If (3.11) is violated, then this pair of observables is served by two different Kolmogorov spaces (for this context).

Remark 3.1 This impossibility for some pairs of observables, say a and b, to embed in a single Kolmogorov probability space probabilities with respect to some context, say C, and probabilities with respect to selection contexts is the main reason for invention of the Växjö model.

Independent observables (in the contextual framework) are considered in Section 12.1.3.

3.1.5 Interpretations of Contextual Probabilities

Mathematical probability, i.e., the collection of probability distributions π of a contextual probability space $\mathcal{P}_{\mathrm{cont}} = (\mathcal{C}, \mathcal{O}, \pi)$, see Definition 3.1, can be interpreted in various ways, see [161]. One of the most useful for applications is the *frequency interpretation* of probability. In von Mises' model[4] it is expressed in the form of the *principle of statistical stabilization of relative frequencies,* $v_C^a(\alpha; N) = n_C^a(\alpha)/N$, for results of observations. Here $n_C^a(\alpha)$ is the number of observations under context C with the fixed result $a = \alpha$. It is postulated that these frequencies stabilize when the number of observations $N \to \infty$. Their limiting value is called the probability of realization of the value α of the observable a:

$$\mathbf{P}(a = \alpha | C) = \lim_{N \to \infty} v_C^a(\alpha; N). \tag{3.12}$$

In the majority of considerations in this book contextual probabilities are interpreted as frequency probabilities. It is the most general interpretation of contextual probability. In principle, one can proceed with this interpretation without considering systems. Measurement is performed under context C, the end of the story! Of course, context C should be repeatable. It should be reproducible sufficiently many times (in the limit – infinitely many times).

We remark that in Kolmogorov's measure-theoretic model the frequency interpretation is expressed in the form of the *law of large numbers.* However, there are some interpretational complications; von Mises strongly criticized attempts to use the law of large numbers as a basis of the frequency interpretation.

Probabilities can also be interpreted as ensemble probabilities – proportions of various results of measurements in sufficiently big ensembles.[5] This definition works especially well for finite sample spaces

$$\mathbf{P}(a = \alpha | C) = n_C^a(\alpha)/N, \tag{3.13}$$

where N is the number of elements in the finite ensemble S_C representing context C and $n_C^a(\alpha)$ is the number of observations with the result $a = \alpha$ in this ensemble.

Another popular interpretation of probability is the *subjective interpretation.* The probability of an event A is a measure of personal belief in realization A. Contextual probability can be interpreted as well as subjective probability, see especially Chap. 7.

3.2 Formula of Total Probability with Interference Term

Let $M = (\mathcal{C}, \mathcal{O}, \mathcal{D}(\mathcal{O}, \mathcal{C}))$ be a Växjö model such that the set of observables $\mathcal{O} = \{a, b\}$ and a, b are dichotomous observables. Let $C \in \mathcal{C}$. There are no reasons to assume that all probability distributions in $D(a, b, C)$ should be described by a single Kolmogorov probability space $\mathcal{P} = (\Omega, \mathcal{F}, \mathbf{P})$. Thus the classical

---

[4] This is the basic frequency probability model [299], see [161] for the modern presentation of von Mises' approach, its generalization and applications to QM.

[5] See [161] for an attempt to generalize this definition to infinite samples by using so-called p-adic probability.

(Kolmogorovian) formula of total probability (FTP), Sect. 2.1.3, can be violated. In general,

$$\mathbf{P}(b = \beta | C) \equiv p_C^b(\beta) \neq \sum_\alpha p_C^a(\alpha) p_{\beta|\alpha}. \tag{3.14}$$

Thus it is impossible to predict the probability of the result $b = \beta$ on the basis of probabilities of results of a-measurements and transition probabilities.

The difference between the left-hand and right-hand sides, denoted by $\delta(\beta | a, C)$, provides a probabilistic measure of $b|a$-*interference* with respect to context C. We will see in Chap. 4 that this coefficient can be really interpreted in terms of interference of "waves of probability." Directly by definition of $\delta(\beta | a, C)$ we can write an equation that is similar to the classical FTP (2.23)

$$p_C^b(\beta) = \sum_\alpha p_C^a(\alpha) p_{\beta|\alpha} + \delta(\beta | a, C). \tag{3.15}$$

This formula has the same structure as the quantum formula of total probability (2.59): [classical part] + additional term, cf. (2.59). We will use in future, Chapter 7, the following simple fact:

$$\sum_{\beta \in X_b} \delta(\beta | a, C) = 0. \tag{3.16}$$

To write the additional term in the same form as in the quantum representation of probabilistic data, we perform the normalization of the probabilistic measure of interference by the square root of the product of all probabilities

$$\lambda(\beta | a, C) = \frac{\delta(\beta | a, C)}{2\sqrt{\prod_\alpha p^a(\alpha) p_{\beta|\alpha}}}. \tag{3.17}$$

The coefficient $\lambda(\beta | a, C)$ also will be called the probabilistic measure of interference. By using this coefficient we can rewrite (3.15) in the QL form:

$$p_C^b(\beta) = \sum_\alpha p_C^a(\alpha) p_{\beta|\alpha} + 2\lambda(\beta | a, C) \sqrt{\prod_\alpha p_C^a(\alpha) p_{\beta|\alpha}}. \tag{3.18}$$

The coefficient of interference $\lambda(\beta | a, C)$ is well defined only in the case when all probabilities $p_C^a(\alpha)$, $p_{\beta|\alpha}$ are strictly positive.

A context C is said to be *a-nondegenerate* if

$$p_C^a(\alpha) \neq 0 \tag{3.19}$$

for any value α of a. We remark that, for $\beta \in X_b$, context C_β is a-nondegenerate if

$$p_{\alpha|\beta} \equiv \mathbf{P}(a = \alpha|C_\beta) \neq 0, \quad \alpha \in X_a. \tag{3.20}$$

A b-nondegenerate context is defined in the same way. We remark that, for $\alpha \in X_a$, context C_α is b-nondegenerate if

$$p_{\beta|\alpha} \neq 0, \quad \beta \in X_b. \tag{3.21}$$

By considering $a|b$-conditioning, instead of $b|a$-conditioning, similarly to (3.18) we have

$$p_C^a(\alpha) = \sum_\beta p_C^b(\beta) p_{\alpha|\beta} + 2\lambda(\alpha|b, C) \sqrt{\prod_\beta p_C^b(\beta) p_{\alpha|\beta}}. \tag{3.22}$$

Definition 3.4 Observables a and b are called supplementary if (3.20) and (3.21) hold.

Remark 3.2 (Complementarity, supplementarity, incompatibility). Condition (3.20), $\mathbf{P}(a = \alpha|C_\beta) \neq 0, \ \alpha \in X_a$, is equivalent to condition $\mathbf{P}(a = \alpha|C_\beta) \neq 1, \ \alpha \in X_a$. Thus it is impossible to determine a value $a = \alpha$ by fixing the value $b = \beta$ (selection context C_β). The result of b-measurement can never predetermine the result of subsequent a-measurement and vice versa – see (3.21). Thus any measurement of a provides additional, or supplementary, information that has not been produced by preceding measurement of b (and vice versa). I also would like to make a comment on terminology. Of course, it would be much better to call the observables complementary. However, Bohr has already reserved this terminology for observables that are mutually exclusive, or incompatible. They cannot, in principle, be measured simultaneously. Supplementarity does not imply mutual exclusivity (incompatibility). We are comfortable with possibility that a and b can be (but need not be) measured simultaneously.

Theorem 3.1 *Let reference observables be supplementary and let a context $C \in \mathcal{C}$ be both a- and b-nondegenerate. Then QL formulas of total probability (3.18) and (3.22) hold.*

Finally, we remark that in application it may be useful to call the coefficient of interference λ *coefficient of supplementarity.* The latter expresses better the real meaning of this coefficient.

Chapter 4
Quantum-like Representation Algorithm – QLRA

As was pointed out in Section 1.8, starting with FTP with the interference term we can construct the representation of a special class of contexts of the Växjö model, so-called *trigonometric contexts*, in complex Hilbert space. Then we obtain Born's rule and the representation of the reference observables by (noncommutative) self-adjoint operators $\widehat{a}$ and $\widehat{b}$. (Noncommutativity of operators is equivalent to consideration of supplementary reference observables.) If the matrix of transition probabilities is *doubly stochastic,* we obtain the conventional QM. However, if it is not doubly stochastic, our QL formalism is more general than the formalism created by Dirac [90] and von Neumann [301].

We shall present a simple algorithm for transferring the probabilistic data $D(a, b, C)$ collected for context C, see Section 3.1.3, (3.6), into a complex probabilistic amplitude, *quantum-like representation algorithm*, QLRA:

$$C \mapsto D(a, b, C) \mapsto \psi.$$

The main distinguishing feature of QLRA is that classical probabilistic data is coupled with its QL image by Born's rule.

Most amazing is the discovery of a representation of a special class of contexts (probabilistic data), not by complex but by hyperbolic probabilistic amplitudes.[1]

Probabilistic data can be classified by estimation of the interference coefficient λ. If it does not exceed 1, data can be represented in complex Hilbert space, in the opposite case in the hyperbolic one. The mixed hyper–complex representation also can appear for some data. For simplicity we shall not consider it in this book, see [214].

Recently QLRA was realized in the Mathematica-6 setup. It provides a way to simulate and to visualize QLRA's work. For the latter purpose we used representation of complex probability amplitudes (in the two-dimensional case) by vectors

---

[1] Instead of the field of complex numbers $\mathbf{C}$ with elements $z = x + iy, i^2 = -1$, the algebra of hyperbolic numbers is considered, $\mathbf{G} : z = x + jy, j^2 = +1$. Here x, y are real numbers. One can proceed in the hyperbolic framework parallel to the conventional complex Hilbert space representation.

A. Khrennikov, *Ubiquitous Quantum Structure,*
DOI 10.1007/978-3-642-05101-2_4, © Springer-Verlag Berlin Heidelberg 2010

on the Bloch sphere. We created a simulator transforming probabilistic data of any origin into Bloch vectors. The same was done in the hyperbolic case. Here we introduced a hyperbolic analogue of the Bloch sphere, so to say the "Bloch hyperboloid", which is used for QL-visualization.

We remark that reading the sections on hyperbolic representation needs more algebraic intuition. In practically all applications we proceed with the complex representation of data. Thus, in principle, the reader can omit Sections 4.5 and 4.6.

The main presentation in this chapter is done under the *assumption that the matrix of transition probabilities* $\mathbf{P}^{b|a}$ *is doubly stochastic* – **DS**. In fact, QLRA works well even for non-doubly stochastic matrices, see Remark 4.2. However, the structure of representation in, e.g., complex Hilbert space is essentially more complicated than in the case of double stochasticity. Unlike the latter case, it is impossible to represent both observables a and b by self-adjoint operators (symmetric matrices): only one of them can be in general given by a self-adjoint operator, the other can be non-self-adjoint. Thus, departure from the case of doubly stochastic matrices induces generalization of the conventional (Dirac–von Neumann, see Chapter 2) quantum formalism. We emphasize that $\mathbf{P}^{b|a}$ matrices which arise in applications, e.g., to psychology, are typically non-doubly stochastic. Therefore, although QLRA produces representation of data in Hilbert space, it is not the conventional quantum representation (even in the complex case), see Section 4.4 for details; see Section 12.5 for the generalized quantum formalism.

In fact, even the condition **DS** is not sufficient to obtain complete consistency with conventional QM formalism. Under this condition we only mimic the most essential features of QM: Born's rule for both reference observables a and b and representation of them by self-adjoint operators. The tricky point is that even if both matrices of transition probabilities $\mathbf{P}^{b|a}$ and $\mathbf{P}^{a|b}$ are doubly stochastic, then QLRA produces two in general nonequivalent (in the sense of unitary equivalence) representations. They are equivalent only if observables are symmetrically conditioned, i.e., condition (2.18) holds. I do not expect the appearance of such matrices of transition probabilities – contextual probabilities for successive measurements – in, e.g., psychology or economics. I suspect that in QM they appear as a trace of existence of rotationally invariant physical space. It seems that "mental space" does not have such a homogeneous structure, see [159, 181].

4.1 Inversion of Born's Rule

We recall that Born's rule (2.32) is an algorithm to transfer complex amplitudes (or in the Hilbert space formalism – normalized vectors) to probabilities. This rule was postulated by Max Born, see Remark 2.1. In fact, this chapter is devoted to the "inverse Born's rule problem:"

IBP (inverse Born problem): *To construct a representation of probabilistic data by complex probability amplitudes that match Born's rule.*

Solution of IBP would provide a way to represent probabilistic data by "wave functions" and operate with this data using linear algebra (as we do in conventional QM). In particular, one would be able to find quantum-like (QL) effects in data collected in any domain of science, e.g., interference of probabilities. We consider the simplest situation.

Let $M = (\mathcal{C}, \mathcal{O}, \mathcal{D}(\mathcal{O}, \mathcal{C}))$ be a contextual statistical model such that $\mathcal{O} = \{a, b\}$. These observables are dichotomous: $a = \alpha_1, \alpha_2$ and $b = \beta_1, \beta_2$. They can be physical (classical or quantum) observables or e.g. two questions that are used for tests in psychology or cognitive or social science and so on. As usual, $X_a = \{\alpha_1, \alpha_2\}$ and $X_b = \{\beta_1, \beta_2\}$, the "spectra of observables". We assume that these observables are supplementary, see Definition 3.4.

We recall that, for each context $C \in \mathcal{C}$, the data $D(a, b, C)$ contain the matrix of transition probabilities $\mathbf{P}^{b|a} = (p_{\beta|\alpha})$. There are also given probabilities $p_C^a(\alpha)$, $\alpha \in X_a$, and $p_C^a(\beta)$, $\beta \in X_b$, see (3.1).

Our aim is to represent this data by a probability amplitude $\psi \equiv \psi_C$ (in the simplest case it is complex valued) such that Born's rule holds for both observables:

$$p_C^b(\beta) = |\langle \psi, e_\beta^b \rangle|^2 , \qquad p_C^a(\alpha) = |\langle \psi, e_\alpha^a \rangle|^2 , \tag{4.1}$$

where $\{e_\beta^b\}_{\beta \in X_b}$ and $\{e_\alpha^a\}_{\alpha \in X_a}$ are orthonormal bases (which are also produced by QLRA) for observables b and a, respectively. These observables are represented by operators $\widehat{b}$ and $\widehat{a}$, which are diagonal in these bases.

4.2 QLRA: Complex Representation

In Section 3.2 we derived the following formula for interference of probabilities:

$$p_C^b(\beta) = \sum_\alpha p_C^a(\alpha) p_{\beta|\alpha} + 2\lambda_\beta \sqrt{\prod_\alpha p_C^a(\alpha) p_{\beta|\alpha}} , \tag{4.2}$$

where the *coefficient of interference* (supplementarity) is

$$\lambda_\beta \equiv \lambda(\beta|a, C) = \frac{p_C^b(\beta) - \sum_\alpha p_C^a(\alpha) p_{\beta|\alpha}}{2\sqrt{\prod_\alpha p_C^a(\alpha) p_{\beta|\alpha}}} . \tag{4.3}$$

To simplify considerations, we shall proceed under the conditions

DS: The matrix of transition probabilities $\mathbf{P}^{b|a}$ is doubly stochastic. (All entries are nonnegative and all rows and all columns sum to 1.)

PO: Probabilistic data $D(a, b, C)$ consist of strictly positive probabilities, i.e., context C is both a- and b-nondegenerate.[2]

We proceed under the following basic assumption (specifying the type of representation):

RC: Coefficients of interference λ_β, $\beta \in X_b$, are bounded by 1:

$$|\lambda_\beta| \leq 1.$$

Probabilistic data $D(a, b, C)$ or simply a context C such that **RC** holds is called *trigonometric*, because in this case we have the conventional formula of trigonometric interference:

$$p_C^b(\beta) = \sum_\alpha p_C^a(\alpha) p_{\beta|\alpha} + 2 \cos \phi_\beta \sqrt{\prod_\alpha p_C^a(\alpha) p_{\beta|\alpha}}, \qquad (4.4)$$

where $\lambda_\beta = \cos \phi_\beta$. This is simply a new parametrization: a new parameter ϕ is used, instead of λ. Parameters ϕ_β are said to be $b|a$-*relative phases* for context C. We defined these phases purely on the basis of probabilities.[3] We denote the collection of *trigonometric contexts* by the symbol C^{tr}. By using the elementary formula

$$D = A + B + 2\sqrt{AB} \cos \phi = |\sqrt{A} + e^{i\phi}\sqrt{B}|^2,$$

for real numbers $A, B > 0$, $\phi \in [0, 2\pi]$, we can represent the probability $p_C^b(\beta)$ as the square of the complex amplitude (Born's rule)

$$p_C^b(\beta) = |\psi(\beta)|^2. \qquad (4.5)$$

Here

$$\psi(\beta) \equiv \psi_C(\beta) = \sqrt{p_C^a(\alpha_1) p_{\beta|\alpha_1}} + e^{i\phi_\beta}\sqrt{p_C^a(\alpha_2) p_{\beta|\alpha_2}}, \quad \beta \in X_b. \qquad (4.6)$$

The formula (4.6) gives the QLRA. For any trigonometric context C, QLRA produces the complex amplitude ψ. This algorithm can be used in any domain of science to create the QL representation of probabilistic data.

---

[2] We recall that from the very beginning observables a and b were chosen supplementary. Thus elements of matrices of transition probabilities $\mathbf{P}^{b|a}$ and $\mathbf{P}^{a|b}$ are strictly positive.

[3] We did not start with a linear space; in contrast, we define geometry from probability. In the conventional quantum formalism, the formula of interference of probabilities is derived starting directly with the Hilbert space. We recall that in QM interference of probabilities is derived via transition from the basis for the a-observable to the basis for the b-observable. From the very beginning observables are given by self-adjoint operators.

We denote the space of functions $\psi : X_b \to \mathbf{C}$ by the symbol $\Phi = \Phi(X_b, \mathbf{C})$. Since $X_b = \{\beta_1, \beta_2\}$, Φ is the two-dimensional complex linear space. By using QLRA we construct the map $J^{b|a} : C^{\text{tr}} \to \Phi(X, \mathbf{C})$, which maps probabilistic data into complex amplitudes. The representation (4.5) of probability is nothing but the famous *Born rule*. The complex amplitude $\psi(\beta)$ can be called a *wave function* of context C or data $D(a, b, C)$.

By using the terminology of quantum information theory we can say that QLRA represents probabilistic data (of a special sort, namely, trigonometric) by *qubits*.

We set $e_\beta^b(x) = \delta(\beta - x)$, $x \in X_b$, Dirac delta-functions concentrated in points $\beta = \beta_1, \beta_2$. Born's rule for complex amplitudes (4.5) can be rewritten in the following form:

$$p_C^b(\beta) = |\langle \psi, e_\beta^b \rangle|^2,$$

where the scalar product in the space $\Phi(X_b, \mathbf{C})$ is defined by the standard formula

$$\langle \psi_1, \psi_2 \rangle = \sum_{\beta \in X_b} \psi_1(\beta)\overline{\psi_2(\beta)}. \tag{4.7}$$

The system of functions $\{e_\beta^b\}_{\beta \in X_b}$ is an orthonormal basis in the Hilbert space $\mathcal{H} = (\Phi, \langle \cdot, \cdot \rangle)$:

$$e_{\beta_1}^a = \begin{pmatrix} 1 \\ 0 \end{pmatrix}, \qquad e_{\beta_2}^a = \begin{pmatrix} 0 \\ 1 \end{pmatrix}. \tag{4.8}$$

Now let $X_b \subset \mathbf{R}$ (in general β is just a label for a result of observation). By using the Hilbert space representation of Born's rule we obtain the Hilbert space representation of the expectation of the observable b,

$$E[b|C] = \sum_{\beta \in X_b} \beta |\psi_C(\beta)|^2 = \sum_{\beta \in X_b} \beta \langle \psi_C, e_\beta^b \rangle \overline{\langle \psi_C, e_\beta^b \rangle} = \langle \hat{b}\psi_C, \psi_C \rangle, \tag{4.9}$$

where the (self-adjoint) operator $\hat{b} : \mathcal{H} \to \mathcal{H}$ is determined by its eigenvectors: $\hat{b}e_\beta^b = \beta e_\beta^b$, $\beta \in X_b$. This is the multiplication operator in the space of complex functions $\Phi(X_b, \mathbf{C}) : \hat{b}\psi(\beta) = \beta\psi(\beta)$. It is natural to represent the b-observable (in the Hilbert space model) by the diagonal operator

$$\hat{b} = \begin{pmatrix} \beta_1 & 0 \\ 0 & \beta_2 \end{pmatrix} \tag{4.10}$$

Remark 4.1 All previous considerations can be applied to the non-doubly stochastic matrix of transition probabilities $\mathbf{P}^{b|a}$. In particular, the basic rule (4.6) determining QLRA can be applied without double stochasticity, as well as (4.5), (4.9), (4.10).

To solve IBP completely, we would like to have Born's rule not only for the b-variable, but also for the a-variable: $p_C^a(\alpha) = |\langle \psi, e_\alpha^a \rangle|^2, \alpha \in X_a$. How can we define the basis $\{e_\alpha^a\}$ corresponding to the a-observable? Such a basis can be found by starting with interference of probabilities. We have

$$\psi = \sqrt{p_C^a(\alpha_1)} f_{\alpha_1}^a + \sqrt{p_C^a(\alpha_2)} f_{\alpha_2}^a , \tag{4.11}$$

where

$$f_{\alpha_1}^a = \begin{pmatrix} \sqrt{p_{\beta_1|\alpha_1}} \\ \sqrt{p_{\beta_2|\alpha_1}} \end{pmatrix} , \qquad f_{\alpha_2}^a = \begin{pmatrix} e^{i\phi_{\beta_1}} \sqrt{p_{\beta_1|\alpha_2}} \\ e^{i\phi_{\beta_2}} \sqrt{p_{\beta_2|\alpha_2}} \end{pmatrix} . \tag{4.12}$$

The condition **DS** implies that the system of vectors $\{f_{\alpha_i}^a\}$ is an orthonormal basis iff the probabilistic phases satisfy the constraint (see [185, 186, 214])

$$\phi_{\beta_2} - \phi_{\beta_1} = \pi \bmod 2\pi , \tag{4.13}$$

i.e., the phases cannot be chosen independently. Thus, instead of the a-basis (4.12), which depends on phases, we can consider a new a-basis that depends only on the matrix of transition probabilities $\mathbf{P}^{b|a}$

$$e_{\alpha_1}^a = \begin{pmatrix} \sqrt{p_{\beta_1|\alpha_1}} \\ \sqrt{p_{\beta_2|\alpha_1}} \end{pmatrix} , \qquad e_{\alpha_2}^a = \begin{pmatrix} \sqrt{p_{\beta_1|\alpha_2}} \\ -\sqrt{p_{\beta_2|\alpha_2}} \end{pmatrix} . \tag{4.14}$$

In this basis ψ is represented as

$$\psi = \sqrt{p_C^a(\alpha_1)} e_{\alpha_1} + e^{i\phi_{\beta_1}} \sqrt{p_C^a(\alpha_2)} e_{\alpha_2}. \tag{4.15}$$

The a-observable is represented by the operator $\hat{a}$, which is diagonal with eigenvalues α_1, α_2 in the basis $\{e_\alpha^a\}$. The average of the observable a coincides with the quantum Hilbert space average:

$$E[a|C] = \sum_{\alpha \in X_a} \alpha p_C^a(\alpha) = \langle \hat{a}\psi_C, \psi_C \rangle. \tag{4.16}$$

We remark that the matrix of transition probabilities $\mathbf{P}^{b|a}$ was assumed to be doubly stochastic. Thus

$$e_{\alpha_1}^a = \begin{pmatrix} \sqrt{p} \\ \sqrt{1-p} \end{pmatrix} , \qquad e_{\alpha_2}^a = \begin{pmatrix} \sqrt{1-p} \\ -\sqrt{p} \end{pmatrix} . \tag{4.17}$$

Here $p = p_{\beta_1|\alpha_1} = p_{\beta_2|\alpha_2}$. In the basis $\{e_\beta^b\}$ the operator $\widehat{a}$ is represented by the matrix

$$\widehat{a} = \begin{pmatrix} \alpha_1 p + \alpha_2(1-p) & (\alpha_1 - \alpha_2)\sqrt{p(1-p)} \\ (\alpha_1 - \alpha_2)\sqrt{p(1-p)} & \alpha_1(1-p) + \alpha_2 p \end{pmatrix}. \tag{4.18}$$

In our approach only observables a and b (e.g., given in the form of questions) were given from the very beginning. The matrix representation of these observables, $a \to \widehat{a}, b \to \widehat{b}$, was constructed on the purely probabilistic basis (by using QLRA).

Suppose now that we have a third observable, say c. We can couple it either with a or with b. In the first case we proceed with QLRA for the pair of reference observables (c, a) and in the second case (c, b). We obtain two representations in complex Hilbert space. Question: *Under which conditions are they unitary equivalent?* Answer: *If all possible pairs of observables are symmetrically conditioned.* It happens in QM, but I am sceptical that it may happen in, e.g., psychology.

4.3 Visualization on Bloch's Sphere

Come back to conventional QM, Section 2.3, and consider two-dimensional complex Hilbert space $\mathcal{H}_2 = \mathbf{C} \times \mathbf{C}$. In *quantum information theory* it describes one qubit. This model can be illustrated geometrically by using the so-called *Bloch's sphere* – the unit sphere in three-dimensional real space $\mathbf{R}^3$ given by the equation: $x^2 + y^2 + z^2 = 1$. It is possible to represent vectors of $\mathcal{H}_2$ by points on this sphere. In this way one can cover the whole sphere. The algorithm of this representation is very simple. Consider in $\mathcal{H}_2$ a basis, say $|0\rangle, |1\rangle$. As is common in quantum information theory, we use Dirac's notation. However, a mathematically educated reader can relax: it is just an arbitrary basis in $\mathcal{H}_2$. Take a vector $\psi \in \mathcal{H}_2$ of the form

$$\psi = \cos\theta|0\rangle + \sin\theta e^{i\phi}|1\rangle. \tag{4.19}$$

It is mapped to the point on Bloch's sphere given by spherical coordinates

$$x = \sin 2\theta \cos\phi, \quad y = \sin 2\theta \sin\phi, \quad z = \cos 2\theta.$$

We remark that up to complex factors $c = e^{ik}, k \in [0, 2\pi]$, any vector of the unit sphere S of $\mathcal{H}_2$ can be represented in the form (4.19). Since a pure quantum state is defined up to such a factor, all states are mapped to Bloch's sphere. In other words, the set $\widetilde{S}$, see Section 2.3, of equivalent classes of the unit sphere S is isomorphic to Bloch's sphere.

We now combine QLRA and "Bloch's representation algorithm". We recall that at the moment we work under condition **RC** – the coefficients of interference are bounded by 1. First of all our computer program checks this condition. If **RC** is violated, then the program's output is empty — no point on Bloch's sphere. It is convenient to use the QLRA output in the a-basis. Thus we make the identification

$$|0\rangle = e^a_{\alpha_1}, \quad |1\rangle = e^a_{\alpha_2}.$$

We have $p^a_C(\alpha_1) = \cos^2\theta$, $p^a_C(\alpha_2) = \sin^2\theta$; $\lambda_{\beta_1} = \cos\phi$, $\sin\phi = \pm\sqrt{1 - \lambda^2_{\beta_1}}$. Finally,

$$x = 2\sqrt{p^a_C(\alpha_1)p^a_C(\alpha_2)}\lambda_{\beta_1},$$
$$y = \pm 2\sqrt{p^a_C(\alpha_1)p^a_C(\alpha_2)}\sqrt{1 - \lambda^2_{\beta_1}},$$
$$z = p^a_C(\alpha_1) - p^a_C(\alpha_2).$$

In the computer program we make the parametrization of probabilities $p^a_C(\alpha_1) = q$, $p^a_C(\alpha_2) = 1 - q$, $p^b_{\beta_1} = p$, $p^b_{\beta_1} = 1 - p$. Since we proceed under condition **DS**, the elements of the matrix of of transition probabilities $\mathbf{P}^{b|a}$ can be parametrized as $p_{\beta_1|\alpha_1} = p_{\beta_2|\alpha_2} = P$, $p_{\beta_1|\alpha_2} = p_{\beta_2|\alpha_1} = 1 - P$.

We see from Figs. 4.1 and 4.2 that Bloch's sphere is sufficiently densely covered by vectors (corresponding to complex probability amplitudes produced by QLRA).

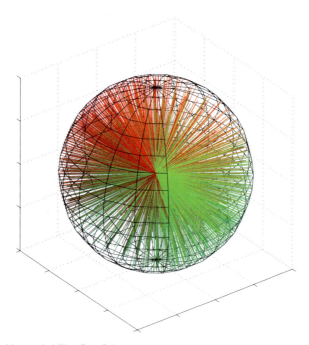

Fig. 4.1 Transition probability $P = 0.1$

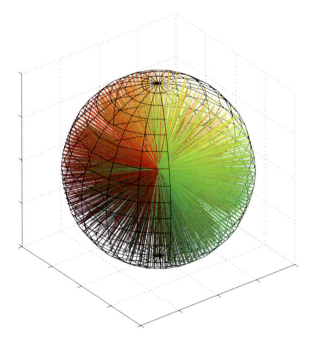

Fig. 4.2 Transition probability $P = 0.5$

4.4 The Case of Non-Doubly Stochastic Matrices

As was already mentioned in Remark 4.1, the b-representation works well even for non-doubly stochastic $\mathbf{P}^{b|a}$. The problem arises when we construct the a-representation. The expansion (4.11) can be written as well: $\psi_C = \sqrt{p_C^a(\alpha_1)} f_{\alpha_1}^a + \sqrt{p_C^a(\alpha_2)} f_{\alpha_2}^a$. However, unlike the case of double stochasticity, we cannot move to the basis $\{e_\alpha^a\}$ given by (4.14). We should try to proceed with the basis $\{f_\alpha^a\}$. The crucial difference is that the second vector of this basis depends irreducibly on context C via phases $\phi_\beta \equiv \phi(\beta|a, C)$. As was shown in [214], it is impossible to find a shift γ such that $\phi_{\beta_1} - \phi_{\beta_2} = \gamma$, mod 2π, for any nontrivial family of contexts. Thus $f^a = f^{a;C}$ and nothing can be done. In this case it is not sufficient to construct the map $J^{b|a} : C^{\mathrm{tr}} \mapsto \mathcal{H}$, mapping contexts (trigonometric) into complex probability amplitudes. This map should be completed by the map $J^{b|a} : C^{\mathrm{tr}} \mapsto \mathcal{E}(\mathcal{H})$, where $\mathcal{E}(\mathcal{H})$ is the space of all (in general nonorthogonal) bases in $\mathcal{H}$ consisting of normalized vectors. Concerning the latter, we remark that, for any α, $\langle f_\alpha^a, f_\alpha^a \rangle = 1$.

Here $C \mapsto f^{a;C}$, where $f_{\alpha_1}^a = \begin{pmatrix} \sqrt{p_{\beta_1|\alpha_1}} \\ \sqrt{p_{\beta_2|\alpha_1}} \end{pmatrix}$, $f_{\alpha_2}^a = \begin{pmatrix} e^{i\phi_{\beta_1}} \sqrt{p_{\beta_1|\alpha_2}} \\ e^{i\phi_{\beta_2}} \sqrt{p_{\beta_2|\alpha_2}} \end{pmatrix}$.

Let us consider a finite-dimensional Hilbert space $\mathcal{H}$. Let $\mathcal{E} = \{e_j\}_{j=1}^n$ be a basis. Then each $\psi \in \mathcal{H}$ can be expanded with respect to this basis

$$\psi = \sum_j u_j e_j, \quad u_j = u_j(\psi) \in \mathbf{C}. \tag{4.20}$$

Denote linear functionals $\psi \to u_\alpha(\psi)$ (determining coefficients) corresponding to the basis $f^{a;C}$ by $u_\alpha^{a;C}$. Then we have

$$p_C^a(\alpha) = |u_\alpha^{a;C}(\psi_C)|^2. \tag{4.21}$$

This is the generalized Born's rule.

Thus in the general case one should operate with nonorthogonal bases to represent observables. One can proceed by using *nonsymmetric operators.* The most natural description is given in the form of *positive operator valued measures* (POVMs), see Section 12.5. However, the use of such generalized quantum observables is not the crucial deviation from the conventional quantum theory. Similar generalized observables are widely used in quantum information theory. In this theory one can also not proceed with conventional Dirac–von Neumann observables given by symmetric operators (orthogonal bases). POVMs are also used there [46, 148]. The only difference is that we should use a more general class of POVM. Unlike [46, 148], our POVM can be nonnormalized, see again Section 12.5. The crucial point is that, unlike QM, context cannot be represented just by a "state", a complex probability amplitude. Even a POVM (and in the simplest case under consideration in this section – basis) depends on context.

In general in applications to cognitive science and psychology matrices of transition probabilities $\mathbf{P}^{b|a}$ are not doubly stochastic, see Chapter 6 and Chapter 7. In this framework one cannot create context-independent representation of both reference observables.

4.5 QLRA: Hyperbolic Representation

4.5.1 Hyperbolic Born's Rule

Instead of the field complex numbers $\mathbf{C}$, we shall use so-called *hyperbolic numbers,* namely, the two-dimensional Clifford algebra, $\mathbf{G}$. We call this algebra *hyperbolic algebra.hyperbolic algebra* Denote by the symbol j the generator of the algebra $\mathbf{G}$ of hyperbolic numbers: $j^2 = 1$. The algebra $\mathbf{G}$ is the two-dimensional (commutative) real algebra with the basis $e_0 = 1$ and $e_1 = j$. Elements of $\mathbf{G}$ have the form $z = x + jy, x, y \in \mathbf{R}$, where $\mathbf{R}$ is the field of real numbers. We introduce an involution in $\mathbf{G}$ by setting $\bar{z} = x - jy$ and set $|z|^2 = z\bar{z} = x^2 - y^2$. We define a hyperbolic exponential function by using a hyperbolic analogue of Euler's formula: $e^{j\theta} = \cosh\theta + j\sinh\theta, \ \theta \in \mathbf{R}$. We remark that $e^{j\theta_1}e^{j\theta_2} = e^{j(\theta_1+\theta_2)}, \overline{e^{j\theta}} = e^{-j\theta}, |e^{j\theta}|^2 = \cosh^2\theta - \sinh^2\theta = 1$. We also have $\cosh\theta = \frac{e^{j\theta}+e^{-j\theta}}{2}, \quad \sinh\theta = \frac{e^{j\theta}-e^{-j\theta}}{2j}$. We shall use the following elementary formula:

$$D = A + B \pm 2AB \cosh\phi = |\sqrt{A} \pm e^{j\phi}\sqrt{B}|^2, \tag{4.22}$$

for real coefficients $A, B > 0$.

We start again with the formula for interference of probabilities (4.2). The condition of double stochasticity is assumed again. However, instead of coefficients of interference that satisfy condition **RC** (which determines the complex representation), we consider coefficients that satisfy the following condition:

RH: Coefficients of interference λ_β, $\beta \in X_b$, are larger than one:

$$|\lambda_\beta| \geq 1.$$

Probabilistic data $D(a, b, C)$ or simply a context C such that **RH** holds is called *hyperbolic*, because in this case the interference of probabilities (4.2) is represented in the hyperbolic form:

$$p_C^b(\beta) = \sum_\alpha p_C^a(\alpha)p_{\beta|\alpha} + 2\varepsilon_\beta \cosh\phi_\beta \sqrt{\prod_\alpha p_C^a(\alpha)p_{\beta|\alpha}}, \tag{4.23}$$

where $\varepsilon_\beta = \text{sign}\,\lambda_\beta$ and

$$|\lambda_\beta| = \cosh\phi_\beta.$$

This is simply a new parametrization: a new parameter ϕ is used, instead of λ. Parameters ϕ_β are said to be hyperbolic $b|a$-*relative phases* for the data C. We remark, see [185, 186, 188, 214], that

$$\varepsilon_{\beta_1} = -\varepsilon_{\beta_2}. \tag{4.24}$$

Thus the interference terms have opposite signs.

We denote the collection of hyperbolic contexts by the symbol $\mathcal{C}^{\text{hyp}}$.

By using (4.23) we can represent the probability p_β^b as the square of the hyperbolic amplitude:

$$p_\beta^b = |\psi(\beta)|^2, \tag{4.25}$$

where

$$\psi(\beta) = \sqrt{p_C^a(\alpha_1)p_{\beta|\alpha_1}} + \varepsilon_\beta e^{j\phi_\beta}\sqrt{p_C^a(\alpha_2)p_{\beta|\alpha_2}}. \tag{4.26}$$

Thus under conditions **DS**, **PO**, and **RH** we represent contexts by the *hyperbolic probability amplitudes*. The domain of application of QLRA has been extended to

cover a new class of contexts (probabilistic data on contexts) – producing coefficients of interference exceeding one. The family of contexts satisfying these conditions is denoted by the symbol C^{hyp} – *hyperbolic contexts*.

In fact, QLRA covers all possible data. As shown in [185, 186, 188, 214], under
condition **DS** any collection of data $D(a, b, C)$ (for supplementary observables) is
either trigonometric or hyperbolic. Thus in this case one could not have one coefficient of interference less than 1 and another larger than 1. But if **DS** is violated, then
contexts can exhibit a mixed *hyper-trigonometric behavior.*

4.5.2 Hyperbolic Hilbert Space Representation

Hyperbolic Hilbert space is **G**-linear space (module) $\mathcal{H}$ with a **G**-linear scalar product: a map $\langle \cdot, \cdot \rangle : \mathcal{H} \times \mathcal{H} \to \mathbf{G}$ that is

1) linear with respect to the first argument:
   $\langle az + bw, u \rangle = a \langle z, u \rangle + b \langle w, u \rangle, \quad a, b \in \mathbf{G}, \quad z, w, u \in \mathcal{H};$
2) symmetric: $\langle z, u \rangle = \overline{\langle u, z \rangle};$
3) nondegenerate: $\langle z, u \rangle = 0$ for all $u \in \mathcal{H}$ iff $z = 0$.

We remark that this generalization of scalar product is not positively defined.

As was pointed out in the introduction to this chapter, in the hyperbolic case we
can proceed in parallel with the complex case. We introduce the space $\Phi(X_b, \mathbf{G})$ of
functions $\psi : X_b \to \mathbf{G}$. Since $X_b = \{\beta_1, \beta_2\}$, $\Phi(X_b, \mathbf{G})$ is the two-dimensional
G-module. We define the **G**-scalar product by

$$\langle \psi_1, \psi_2 \rangle = \sum_{\beta \in X_b} \psi_1(\beta) \overline{\psi_2(\beta)} \tag{4.27}$$

with conjugation in the algebra **G**. We consider hyperbolic Hilbert space $\mathcal{H} =
(\Phi, \langle \cdot, \cdot \rangle)$. Denote by $\{e_\beta^b\}_{\beta \in X_b}$ the orthonormal basis (4.8). Thus we have the hyper-
bolic analogue of Born's rule $p_\beta^b = |\langle \psi, e_\beta^b \rangle|^2$.

Let $X_b \subset \mathbf{R}$. By using Born's rule, we obtain the hyperbolic Hilbert space
representation of the average of the b-observable. Here the (self-adjoint) operator
$\widehat{b} : \mathcal{H} \to \mathcal{H}$ is again determined by its eigenvectors: $\widehat{b} e_\beta^b = \beta e_\beta^b, \beta \in X_b$. This is
the multiplication operator in the space of hyperbolic-valued functions $\Phi(X_b, \mathbf{G})$:
$\widehat{b}\psi(\beta) = \beta \psi(\beta)$. Thus QLRA creates a map from the collection of hyperbolic
contexts into hyperbolic Hilbert space: $J^{b|a} : C^{\mathrm{hyp}} \to \mathcal{H}$.

By generalizing the terminology of quantum information theory we call normalized vectors $\psi \in \mathcal{H}$ *hyperbolic qubits*.

Thus $J^{b|a}$ maps probabilistic data into hyperbolic qubits. To solve IBP com-
pletely (in the hyperbolic case), we would like to have Born's rule not only for the
b-variable, but also for the a-variable: $p_C^a(\alpha) = |\langle \psi, e_\alpha^a \rangle|^2, \alpha \in X_a$.

How can we define the basis $\{e_\alpha^a\}$ corresponding to the a-observable? Such a basis can be found starting with interference of probabilities. We have for ψ-output of QLRA $\psi = \sqrt{p_C^a(\alpha_1)} f_{\alpha_1}^a + \sqrt{p_C^a(\alpha_2)} f_{\alpha_2}^a$, where

$$f_{\alpha_1}^a = \begin{pmatrix} \sqrt{P_{\beta_1|\alpha_1}} \\ \sqrt{P_{\beta_2|\alpha_1}} \end{pmatrix}, \quad f_{\alpha_2}^a = \begin{pmatrix} \varepsilon_{\beta_1} e^{j\phi_{\beta_1}} \sqrt{P_{\beta_1|\alpha_2}} \\ \varepsilon_{\beta_2} e^{j\phi_{\beta_2}} \sqrt{P_{\beta_2|\alpha_2}} \end{pmatrix}. \tag{4.28}$$

The condition **DS** implies that the system of vectors $\{f_{\alpha_i}^a\}$ is an orthonormal basis iff the probabilistic phases satisfy the constraint $\phi_{\beta_2} = \phi_{\beta_1}$. We also recall that $\varepsilon_{\beta_2} = -\varepsilon_{\beta_1}$. Thus, instead of the a-basis (4.28), which depends on phases and signs, we can consider a new a-basis, which depends only on the matrix of transition probabilities $\mathbf{P}^{b|a} - e_{\alpha_1}^a = \begin{pmatrix} \sqrt{P_{\beta_1|\alpha_1}} \\ \sqrt{P_{\beta_2|\alpha_1}} \end{pmatrix}, \quad e_{\alpha_2}^a = \begin{pmatrix} \sqrt{P_{\beta_1|\alpha_2}} \\ -\sqrt{P_{\beta_2|\alpha_2}} \end{pmatrix}$. In this basis ψ is represented as $\psi = \sqrt{p_C^a(\alpha_1)} e_{\alpha_1}^a + \varepsilon_{\beta_1} e^{j\phi_{\beta_1}} \sqrt{p_C^a(\alpha_2)} e_{\alpha_2}^a$. The a-observable is represented by the operator $\widehat{a}$, which is diagonal with eigenvalues α_1, α_2 in the basis $\{e_\alpha^a\}$. The average of the observable a coincides with the hyperbolic Hilbert space average, see (4.16). In the basis $\{e_\beta^b\}$ this operator is represented by the matrix (4.18). Thus the difference between complex and hyperbolic representations is really minimal.

4.6 Bloch's Hyperboloid

Consider hyperbolic QM, i.e., quantum formalism based on hyperbolic Hilbert space, see [174, 172] for more details. Consider two-dimensional hyperbolic Hilbert space $\mathcal{H}_2 = \mathbf{G} \times \mathbf{G}$. It describes one "hyperbolic qubit." As well as in the complex case, this model can be illustrated geometrically by using what we call *Bloch's hyperboloid* – a hyperboloid in three dimensional real space $\mathbf{R}^3$:

$$x^2 - y^2 + z^2 = 1.$$

It is possible to represent vectors of $\mathcal{H}_2$ by points on this hyperboloid. In this way one can cover the whole hyperboloid. The algorithm of this representation is very simple. Consider in $\mathcal{H}_2$ a basis, say $|0\rangle, |1\rangle$. Take a vector $\psi \in \mathcal{H}_2$ of the form

$$\psi = \cos\theta |0\rangle + \varepsilon \sin\theta e^{j\phi} |1\rangle, \tag{4.29}$$

where $\varepsilon = \varepsilon_{\beta_1}, \phi = \phi_{\beta_1}$. It is mapped to the point on Bloch's hyperboloid given by hyperbolic coordinates

$$x = \varepsilon \sin 2\theta \cosh\phi,$$
$$y = \sinh\phi,$$
$$z = \cos 2\theta \cosh\phi.$$

Fig. 4.3 Transition
probability $P = 0.1$

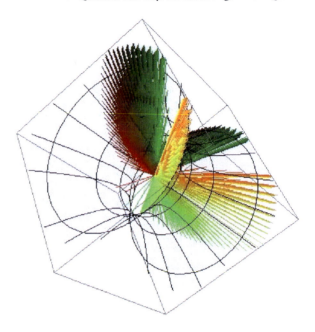

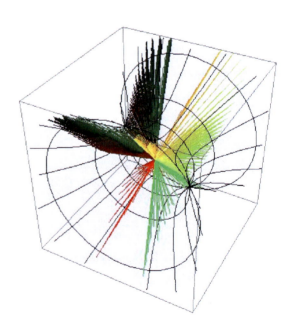

Fig. 4.4 Transition
probability $P = 0.3$

Fig. 4.5 Transition
probability $P = 0.5$

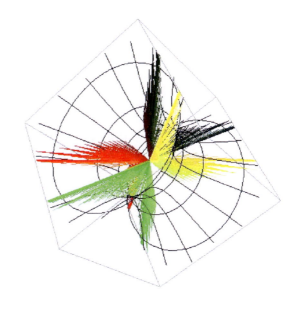

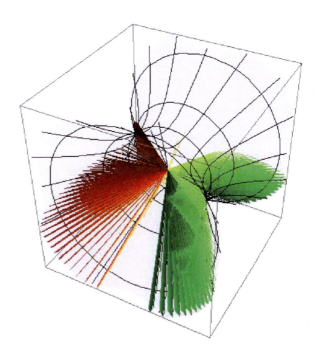

Fig. 4.6 Transition
probability $P = 0.9$

We now combine this "hyperbolic Bloch's algorithm" with QLRA. We recall that we work under condition **RH**. First of all our computer program checks this condition. If **RH** is violated, then the program's output is empty – no point on Bloch's hyperboloid. We make the identification $|0\rangle = e_{\alpha_1}^a$, $|1\rangle = e_{\alpha_2}^a$ and set $p_C^a(\alpha_1) = \cos^2\theta$, $p_C^a(\alpha_2) = \sin^2\theta$. We also have $\lambda_{\beta_1} = \varepsilon_{\beta_1}\cosh\phi_{\beta_1}$. So, we proceed as $|\lambda_{\beta_1}| = \cosh\phi_{\beta_1}$, $\pm\sqrt{\lambda_{\beta_1}^2 - 1} = \sinh\phi_{\beta_1}$. Finally

$$x = 2\varepsilon\sqrt{p_C^a(\alpha_1)p_C^a(\alpha_2)}|\lambda_{\beta_1}|,$$

$$y = \pm\sqrt{\lambda_{\beta_1}^2 - 1},$$

$$z = (p_C^a(\alpha_1) - p_C^a(\alpha_2))|\lambda_{\beta_1}|.$$

In the program we make the same parametrization of probabilities as in the spherical case. Examples of the program's output are given in Fig 4.3–4.6.

Chapter 5
The Quantum-like Brain

In Chapter 3, the contextual probabilistic model was invented: the Växjö model. Now it is applied to the description of mental processes. This description is based on QL representations – by probability amplitudes – in cognitive, social and political sciences, psychology, and economics. In particular, this model suggests interesting cognitive experiments to check QL structures of mental processes. The crucial role is played by interference of probabilities for *mental observables.* Recently, such experiments based on recognition of *ambiguous images* have been performed by Conte et al. [66, 67]. These experiments confirmed my prediction [173, 180] of the QL behavior of mind. In the Växjö approach "quantumness of mind" has no direct relation to the fact that the brain (as any physical body) is composed of quantum particles. A new terminology *quantum-like mind* is used. Cognitive QL behavior is characterized by a nonzero coefficient of interference (supplementarity) λ, see Section 3.2. It can be found on the basis of statistical data. The hypothesis of QL mind can be tested experimentally!

The Växjö model predicted, see Chapter 4, not only $\cos\theta$ interference of probabilities, but also hyperbolic $\cosh\theta$ interference. The latter type of interference has never been observed for physical contexts, but such a possibility cannot be excluded for cognitive systems, see [275] and Chapter 7 for more details.

In this chapter, a model of the brain's functioning as a *QL computer* is proposed; the difference between quantum and QL computers is discussed.

5.1 Quantum and Quantum-like Cognitive Models

The idea that the description of the brain's functioning, cognition, and consciousness cannot be reduced to the theory of neural networks and dynamical systems (see Ashby [19], Hopfield [150], Amit [15], Strogatz [285], van Gelder [296], van Gelder and Port [297]), and that quantum theory may play an important role has been discussed in a huge variety of forms, see, e.g., Whitehead [303] , Orlov [247], Albert and Loewer [10], Albert [11], Healey [138], Lockwood [229], Penrose [249, 250], Donald [91– 93], Jibu and Yasue [154], Bohm and Hiley [40], Stapp [284], Hameroff [128, 129], Loewer [230], Hiley and Pylkkänen [144],

Deutsch [86], Barrett [29], Khrennikov [161, 173, 175, 176, 180, 198], Vitiello [298] and literature therein.

One dominant approach to the application of QM formalism to the description of brain functioning is *quantum reductionism*, see e.g. Hameroff [128, 129] and Penrose [249, 250]. This was a new attempt at physical reduction of mental processes, cf. Ashby [19], Hopfield [150], Amit [15]. This is an interesting project of great complexity and it is too early to draw any conclusions about its future. One important contribution of quantum reductionism is critique of the classical reductionist approach (neural networks and dynamical systems approach) and artificial intelligence, see especially Penrose [249, 250]. On the other hand, quantum reductionism has been strongly criticized by neurophysiologists and cognitive scientists, who assume that the *neuron is the basic unit of processing of mental information*.

We mention the *quantum logic* approach: mind cannot be described by classical logic and the formalism of quantum logic should be applied. Orlov [247] published the first paper in which this idea was explored. It is important to remark that he discussed interference within a single mind. Such an interference was also discussed by Deutsch [86]. We point to extended investigations based on the *many-minds* approach, see Healey [138], Albert and Loewer [10], Albert [11], Lockwood [229], Donald [91– 93], Loewer [230], Barrett [29], etc. Finally, we mention attempts to apply *Bohmian mechanics* to describe mental processes – Bohm and Hiley [40], Hiley and Pylkkänen [144], Khrennikov [161], Choustova [54–62].

In [198] I developed the theory of "quantum-like mind", which is presented in this chapter.[1] As was already emphasized, the QL approach has nothing to do with *quantum reductionism*. Of course, I do not claim that my approach implies that quantum physical reduction of mind is totally impossible. However, I can explain the main QL feature of mind – *interference of minds* – without reduction of mental processes to quantum physical processes. Consequently my QL model does not face such horrible problems of QM as nonlocality or death of realism.[2] One may ask:

Why is it so important to combine realism with quantum probabilistic features in neurophysiology, cognitive sciences, psychology and sociology?

A fundamental consequence of the possibility of such a combination is that macroscopic neuronal structures (in particular, a single neuron) as well as cognitive and psychological contexts can exhibit QL features. It is possible to eliminate the fundamental problem disturbing adherents of quantum physical reductionism:

How can one combine the neuronal (macroscopic) and quantum (microscopic) models?

---

[1] Recently, Busemeyer, a professor of psychology, has explained some paradoxical features of psychological behavior by using a QL model that was based on an approach very similar to that developed in the author's papers, namely, on the QL deformation of the classical formula of total probability, see e.g. [48, 49]. It is amazing that people working in such different domains of science as foundations of probability theory and psychology arrive at similar models.

[2] I reject the idea of using quantum nonlocality in cognitive science as totally absurd.

It is a terrible problem for everybody who tries to proceed with quantum reductionism, e.g., for Penrose [250]: "It is hard to see how one could usefully consider a quantum superposition consisting of one neuron *firing,* and simultaneously *non-firing.*"

In the Växjö model it is possible to operate with QL probabilities without appealing to such a notion as *superposition of states of a single system,* see Chapter 4. All distinguishing probabilistic features of quantum mechanics can be obtained without it. This implies that, unlike quantum reductionism, there is no need to look for the microscopic basis of mental processes.[3] In my model *"mental interference"* is not based on superposition of individual quantum states. Mental interference is described in a classical (but contextual) probabilistic framework. A *mental wave function* represents not a mental state of an individual cognitive system, but a neurophysiological, cognitive or psychological context C, see Chapter 4.[4]

In particular, Växjö model can be applied to the description of mental observations in the QL terms. We start with *mental interference,* which is defined as interference of probability distributions of two *supplementary mental observables,* see Definition 3.2, Section 3.2. For example, in psychology such observables can be realized in the form of two supplementary questions that are asked to people participating in a test. A condition of supplementarity can be checked easily on the basis of experimental statistical data collected in the form of "yes-no" answers to questions. The magnitude of mental interference is characterized by a coefficient of interference (or supplementarity) λ. Depending on this magnitude we obtain different representations of probabilities in experiments with cognitive systems. In particular, we obtain the QL representation of cognitive (or social, or economic) contexts in complex (or maybe even hyperbolic) Hilbert space, by using the representation algorithm, QLRA, given in Chapter 4. This approach should be justified experimentally. A priori there is no reason for cognitive systems to exhibit QL probabilistic behavior, in particular, nontrivial interference. We present the detailed description of a few experimental tests to check the hypothesis of QL probabilistic behavior. We hope that a variety of such tests will be performed in various domains of science: psychology, cognitive science and sociology, economics, see [66, 67].

---

[3] Reductionists should do this and go to the deepest scales of space and time to find some reasonable explanation of superposition and interference (e.g., go inside *microtubules* or to scales of *quantum gravity*).

[4] My comparison of the contextual approach and quantum reductionism cannot be used as an argument against the latter. One could not exclude the possibility that mental processes could be reduced to quantum physical processes, e.g., in microtubules, or that the act of consciousness is really induced by the collapse of the wave function of superposition of two mass states. However, the Växjö model makes it possible to use quantum mathematical formalism in neurophysiology, cognitive science, psychology, and sociology without all those tricky (quantum physical) things that are so important in the reductionist approach.

5.2 Interference of Minds

5.2.1 Cognitive and Social Contexts; Observables

We consider examples of cognitive contexts and observables that can be measured for these contexts.

1) C is a procedure of selection of a specific group S_C of people or animals (creation of an ensemble of cognitive systems). Context C is represented by this group S_C. For example, a group $S_{prof.math.}$ of professors of mathematics is selected. Then one can perform "mental measurements" by asking questions or giving tasks. In the simplest experiment, to check interference of minds, two questions, say a and b, are asked. We can select a group of people of a particular age or a group of people having a specific mental state: for example, people in love, hungry or depressed.

2) C is a learning procedure that is used to create some specific group of people or animals. For example, rats are trained to react to a special stimulus. Students are trained in probability theory. Here a and b are two supplementary questions (Definition 3.4) given in the exam. For instance, a is a theoretical question or task, e.g., to prove the central limit theorem (CLT), and b is a practical question, e.g., to find the average with respect to a given probability distribution.[5] In this example, post-measurement condition (3.5) holds (Section 3.1.1: "projection postulate"). Suppose that a student proved CLT. Ask him to do the same, within a reasonable period of time. We can be practically sure (up to small statistical deviations) that he will prove it again. Thus

$$\mathbf{P}(\text{to prove CLT}|\text{CLT was proven}) = 1$$

as well as $\mathbf{P}(\text{not prove CLT}|\text{CLT was not proven}) = 1$.

3) C is a collection of paintings, C_{paint} (e.g., the collection of the Hermitage in St. Petersburg) and people interact with C_{paint} by looking at the pictures. Mental measurements are based on questions which those people are asked about this collection.

4) C is "context of classical music", $C_{clmus.}$, and people interact with $C_{clmus.}$ by listening to this music. In principle, we need not use an ensemble of different people. It can be one person of whom we ask questions each time after he has listened to a CD (or radio program) of classical music.

The last two examples illustrate why we started with the contextual approach and not simply ensembles of systems. A cognitive context need not be identified with an ensemble of cognitive systems representing this context. For us C_{paint} and $C_{clmus.}$ and not ensembles of people representing them, $S_{C_{paint}}$ and $S_{C_{clmus.}}$, are basic.

---

[5] The problem of supplementarity is very delicate. For example, if the first task was to prove CLT and the second to find the average with respect to a concrete Gaussian distribution, then a and b are definitely not supplementary: $\mathbf{P}(b = +|a = +) = 1$ and $\mathbf{P}(b = -|a = +) = 0$. However, if the first question was on the Poisson distribution and the second to find the average with respect to the Gaussian distribution, then they can be considered as supplementary: $\mathbf{P}(b = \pm|a = \pm) > 0$.

We can also consider *social contexts*, for example, social classes: proletariat and bourgeois contexts; or war and revolution contexts, financial crises context, poverty and welfare contexts, and so on.[6]

5.2.2 Quantum-like Structure of Experimental Mental Data

We describe a mental interference experiment. Let $a \in X_a = \{\alpha_1, \alpha_2\}$ and $b \in X_b = \{\beta_1, \beta_2\}$ be two dichotomous mental observables, e.g., two questions: α_1='yes', α_2='no', β_1='yes', β_2='no'. We use these two fixed reference observables for the probabilistic and then QL representations of cognitive reality given by some context C.[7] This context is assumed to be reproducible such that repeatable measurements of both reference observables can be performed. It can be very sensitive and each measurement may change it essentially.

We perform observations of b under C and obtain frequencies

$$v_C^b(\beta) = \frac{\text{the number of results } b = \beta}{\text{the total number of observations}}, \quad \beta \in X_b.$$

When the total number of observations $N \to \infty$, the frequencies $v_C^b(\beta) \equiv v_C^b(\beta; N)$ approaches the probability $p_C^b(\beta)$ of getting the result β for the b-observation. We also define frequencies $v_C^a(\alpha)$ and probabilities $p_C^a(\alpha)$ for the a-observation.[8]

As was supposed in Section 3.1, selection-contexts are given, e.g., $C_\alpha, \alpha \in X_a$. They are created in the following way. Measurements of a are performed (the question a is asked to all cognitive systems selected for this experiment). Cognitive systems who answered $a = \alpha$ are selected. The C_α produces an ensemble of cognitive systems, say S_{C_α}.[9] Now b-measurements are performed under cognitive context C_α – for the ensemble S_{C_α}. We find frequencies ($\beta \in X_b, \alpha \in X_a$):

$$v_{\beta|\alpha} = \frac{\text{the number of the result } b = \beta \text{ under context } C_\alpha}{\text{the total number of observations under context } C_\alpha},$$

---

[6] The Växjö model can be used in social and political sciences and even in history. One can try to find QL corresponding data. It would be amazing to show that the historical process can exhibit QL features.

[7] In general by choosing another pair of reference observables we shall obtain another representation of cognitive contextual reality. Can we find two fundamental mental observables? This is a very difficult question. In physics the answer is well known: the position and the momentum form the fundamental pair of reference observables. Which mental observables can be chosen as mental analogons of the position and the momentum?

[8] Observables, e.g., questions, should be supplementary, Definition 3.4. The answer, e.g., a ='yes' does not pre-determine (statistically) the answer to the subsequent question b. Moreover, the questions should satisfy post-measurement condition (3.5), Section 3.1.1: "projection postulate".

[9] In the general case the situation is more complicated, Section 5.2.3. However, we restrict considerations to the mentioned scheme of creation of selection contexts. In any event it was used in experiments [66].

and the corresponding probabilities $p_{\beta\alpha}$. Data to find the interference (supplementarity) coefficient (3.17) are collected. We operate with frequencies

$$\lambda_{\mathrm{ex}}(b = \beta | a, C) = \frac{v_C^b(\beta) - v_C^a(\alpha_1)v_{\beta|\alpha_1} - v_C^a(\alpha_2)v_{\beta|\alpha_2}}{2\sqrt{v_C^a(\alpha_1)v_{\beta|\alpha_1} v_C^a(\alpha_2)v_{\beta|\alpha_2}}}. \tag{5.1}$$

An empirical situation with $\lambda_{\mathrm{ex}}(b = \beta | a, C) \neq 0$ would yield evidence for QL behavior of cognitive systems. In this case, starting with the (experimentally calculated) coefficient of interference $\lambda_{\mathrm{exp}}(b = \beta | a, C)$ we can proceed either to the conventional Hilbert space formalism (if this coefficient is bounded by 1) or to so-called hyperbolic Hilbert space formalism (if this coefficient is larger than 1), see Chapter 4 and more in the book [214].

5.2.3 Contextual Redundancy

We remark that in general transition probabilities $p_{\beta\alpha}$ can depend on the original cognitive context C :

$$p_{\beta|\alpha} = p_C(\beta|\alpha)$$

To perform the $[a = \alpha]$-selection, one should first perform measurement of a for some initial context C. In general, there is no reason to hope that after subsequent measurement of another (even supplementary) observable, denoted by b, dependence on C will disappear.

Let us consider the very special case when dependence of the transition probabilities $p_C(\beta|\alpha)$ on C is redundant. For example, students belonging to the group S_C (which was trained under the mental or social conditions C) should answer the question a. After this we select a new ensemble S_{C_α} of students who have answered $a = \alpha$. If this question is so important for a student that he totally forgets about the previous C-training and remembers only the previous answer $a = \alpha$, then the transition probabilities do not depend on C and the index C can be omitted:

$$p_C(\beta|\alpha) \equiv p_{\beta|\alpha}. \tag{5.2}$$

We call (5.2) the condition of *contextual redundancy*. Condition of contextual redundancy is similar to condition of *Markovness* in classical probability theory.

The total destruction of memory of the previous context C (i.e., learning procedure) is too strong a metaphor. It is better to consider an *essential state update*. Thus the memory on C is still present, but the experience generated by an interaction with the question a will dominate in interaction with a subsequent question b. In the example with the exam in probability theory, see Sect. 5.2.1, by proving CLT a student does not destroy the memory of the course in probability theory. However, his state of mind was essentially updated in the process of proving CLT. Consider a number of contexts $C, C', \ldots$ corresponding to courses in probability theory at various universities. Consider $[a = +]$-selection contexts

$$C_+^a(C), C_+^a(C'), \ldots, \tag{5.3}$$

a selection of students who proved CLT in the exams on probability theory. For a sufficiently large spectrum of supplementary questions, the condition of contextual redundancy (5.2) holds. Thus contexts (5.3) can be identified and considered as one context, selection context C_+^a. Of course, condition (5.2) cannot hold for all possible (supplementary) questions b. However, in this book we typically operate only with a pair of supplementary questions.

We remark that contextual redundancy takes place in QM for observables with *nondegenerate spectra*. Here the transition probabilities do not depend on the original context C, the preparation procedure for a quantum state ψ, see formula (2.50), Sect. 2.4. One can (but need not!) also appeal to von Neumann's projection postulate, Sect. 12.3. If quantum observable a is represented by the operator $\widehat{a}$ having nondegenerate spectrum, then the post-measurement state is just one of the eigenvectors of $\widehat{a}$. Memory about the pre-measurement state ψ is completely destroyed by a-measurement. We remark that QM can be considered as a contextual model: contexts are given by quantum states: $C \equiv C_\psi$, see Sect. 12.4. Thus under the condition of contextual redundancy we obtain a class of Växjö models that is the closest to QM (for observables with nondegenerate spectra).

However, we do not want to restrict our considerations to this class of models. How can we proceed in the general case? Some context, say $\Omega \in \mathcal{C}$, should be chosen as a "basic context". Corresponding contexts $C_\alpha^a(\Omega)$ are declared as C_α^a-contexts of the model, cf. with Kolmogorovian contextual models in Sect. 12.4. In the latter case the total space of elementary events Ω is considered as the basic context, and here $C_\alpha^a \equiv C_\alpha^a(\Omega) = \{\omega \in \Omega : a(\omega) = \alpha\}$.

The problem of finding of an adequate basic context $\Omega \in \mathcal{C}$ is very complicated. In fact, transition probabilities encode correlations between observables, see (3.9), Section 3.1.4. Therefore the basic context Ω should be selected to represent the pure (as much as possible) correlation effect between observables a and b. Of course, it depends of the concrete pair of reference observables a and b, i.e., $\Omega = \Omega(a, b)$.

Finally, we come back once again to the example with the exam in probability theory, see Sect. 5.2.1. To prove CLT, a student should invest a lot of effort, in particular, this (very complicated) proof takes time. Thus, as in QM, the process of measurement is a complex process of interaction between a system and a measurement device. In the present example, systems are students, but the a-measurement device is CLT, i.e., a mental structure.[10]

---

[10] Well, there are also teachers in this exam.

5.2.4 Mental Wave Function

The algorithm (QLRA): $C \rightarrow \psi_C$, Chapter 4, represents cognitive, social, psycho-logical, and economic contexts by complex and hyperbolic amplitudes. To obtain a closer analogy with QM, one can speak about the *mental wave function.* One need not imagine "mental waves." In the contextual approach the mental wave function $\psi \equiv \psi_C$ is simply a special representation of probabilistic data collected about context C with the aid of two (specially selected) reference observables a and b.

I speculate that some *cognitive systems developed (in the process of evolution) the ability to operate with mental wave functions,* i.e., to represent probabilistic data in linear space. Roughly speaking, such a system does not feel individual counts, but the general statistics encoded in the ψ_C. In this sense the mental wave function ψ_C is an element of mental reality. Encoding by ψ_C provides a possibility for linear processing of data.

5.3 Quantum-like Projection of Mental Reality

The QL representation for mental processes is a projection of the neuronal model to the complex (or hyperbolic) Hilbert space model. It induces huge loss of information produced by the neurons.

5.3.1 Social Opinion Poll

Let us consider a family of social contexts C such that each context corresponds to the society of some country: C_{USA}, C_{GB}, C_{FR}, ..., C_{GER}, ... and let us consider two reference observables given by the questions

a) "Are you against pollution?" and
b) "Would you like to have lower prices for gasoline?"

It is supposed that observables a and b are supplementary:

$$\mathbf{P}(b = \text{yes}|a = \text{no}) \neq 0, \quad \mathbf{P}(b = \text{no}|a = \text{no}) \neq 0,$$
$$\mathbf{P}(b = \text{yes}|a = \text{yes}) \neq 0, \quad \mathbf{P}(b = \text{no}|a = \text{yes}) \neq 0.$$

Moreover, the transition probabilities $\mathbf{P}(b = \beta|a = \alpha)$ do not depend on a society C, condition of contextual redundance holds. For example, the proportion of people who are against pollution among people who are satisfied by prices for gasoline is the same in the USA, Great Britain, France, and so on. Of course, this is a rather strong assumption.

In our QL-model societies are represented by complex (or maybe hyperbolic?) probability amplitudes

$$\psi_{\text{USA}}, \quad \psi_{\text{GB}}, \quad \psi_{\text{FR}}, \ldots, \quad \psi_{\text{GER}}, \ldots$$

These mental wave functions can be used to describe the dynamics of these societies. However, answers to the questions a and b do not completely characterize a society. Thus this QL representation induces a huge loss of information about the society.

5.3.2 Quantum-like Functioning of Neuronal Structures

Let us consider two coupled neural networks G_1 and G_2. They interact with a family of contexts $\mathcal{C} = \{C\}$, which are given by input signals into both networks. For example, contexts $\mathcal{C} = \{C\}$ can be visual images and networks G_1 and G_2 contribute to recognition of these images, e.g., G_1 is responsible for contours and G_2 for colors. I emphasize from the very beginning that in my model an image in the brain is not created by networks. It is the result of the QL representation of statistics of signals produced by networks.

We use the so-called *frequency-domain approach*, see for example *Hoppensteadt* [151], and assume that cognitive information is presented by frequencies of firing of neurons. We recall that in the process of interaction with the cognitive context frequencies of firing of neurons in, e.g., the network G_1 are synchronized. It is possible to speak about the *"network frequency."* Finally, we point out that each network can be widely distributed in the brain. Thus spatially separated neurons fire synchronously.

Typically a network has a hierarchic structure and the network's frequency can be identified with the frequency of firing of the network's *conductor.* Denote conductors of G_1 and G_2 by symbols c_{G_1} and c_{G_1}. We are aware that the question of the presence of a hierarchic structure of neural networks in the brain and, in particular, the existence of neuron-conductors [14], *"grandmother neurons"*, is still a source of intense debate in the neurophysiological community, see, e.g., [232] on experimental results in favor of the neural hierarchy. Therefore later we will attempt to exclude such conductors from our model. However, the use of them makes the model more illustrative.

Consider two reference observables a, b. Here $a = +$ if the neuron c_{G_1} is firing, and $a = -$ if the neuron c_{G_1} is non-firing, and $b = +$ if the neuron c_{G_2} is firing, and $b = -$ if the neuron c_{G_2} is non-firing. Probabilities $p_C^a(\pm)$, $p_C^b(\pm)$ are defined by frequencies of firing. Consider a possible mechanism of production of frequency probabilities:

Two *time scale* parameters, depending on the cognitive system, are given: Δ is the time scale of production of probabilities ("probabilistic images"), δ is the duration (average) of a pulse from a neuron. Set $\tau = \delta/\Delta$. Let $n_C^a(+)$ be the number of pulses produced by G_1 during the interval Δ (in the process of interaction with cognitive context C). Then probability is given by

$$p_C^a(+) = \tau n_C^a(+), \quad p_C^a(-) = 1 - p_C^a(+).$$

Probabilities $p_C^b(\pm)$ are defined in the same way. These probabilities are easily expressed in networks' frequencies. Let G_1 oscillate (synchronously) with frequency $f_C^{G_1}$ oscillations per second. Then $n_C^a(+) = f_C^{G_1} \Delta$ and

$$p_C^a(+) = f_C^{G_1} \delta. \tag{5.4}$$

Thus it is possible to define probabilities even without involving hierarchic structures and conductor neurons. It is enough to know the frequencies of synchronized (in the process of interaction with C) firings for the corresponding networks. These probabilities provide partial information on the neuronal representation of context C.

Transition probabilities are defined in the following way. First, we should find an appropriate basic context $\Omega = \Omega(G_1, G_2)$, see the very end of Section 5.2.3. As was pointed out, it should be the basis of estimation of pure correlations between two networks G_1 and G_2. So, the specific influence of concrete cognitive context C should be eliminated, as much as possible. One can speculate that Ω corresponds to the state of relaxation. For example, G_1 and G_2, performing the image recognition are not excited by interaction with images.

Denote by $n_{+|+}$ the number of c_{G_2} firings during the periods of c_{G_1} firing, i.e., the number of "matched firings." Then

$$p_{+|+} = \tau n_{+|+}, \quad p_{-|+} = 1 - p_{+|+}.$$

It is also clear how to find probabilities $p_{\pm-}$. Thus the matrix of transition probabilities is created in advance in the state of relaxation.[11]

The brain can now execute QLRA (and we assume that it really can do this) and represent context C (e.g., an image C) by the amplitude ψ_C. This vector in Hilbert space is the mental image of context C.

Of course, ψ_C provides only a rough projection of the neuronal image of the context C. However, we cannot exclude that cognition (and especially consciousness) is really based on such a QL-projecting of neuronal states. The brain makes its decisions by operating with mental wave functions and not with frequencies of firings. In cognitive literature, the problem of the *neural code* is widely discussed. My conjecture is that the neural code is given by QLRA, transforming frequencies of firings into probability amplitudes.

Denote by κ the average time for processing of QLRA, i.e., the time that is required to produce ψ_C on the basis of probabilistic data, namely $W(a, b, C) = \{p_C^a(\pm), p_C^b(\pm)\}$, collected on C. Intervals of time which are less than $\Delta_{\mathrm{cogn}} = \Delta + \kappa$

---

[11] By coupling our model with EEG studies of the brain, we can say that the latter state is characterized by frequencies of α-waves in the brain. The states of active interaction with sufficiently complex cognitive contexts are characterized by frequencies of β- and γ-waves. By (5.4) probabilities $p_C^a(+)$ increase with increasing brain-wave frequencies. In contrast, transition probabilities do not vary; they are rigidly coupled to the α-waves.

has no cognitive meaning. So, the right *cognitive scale* is given by Δ_{cogn}. This scale corresponds to the dynamics of the mental wave function, $t \mapsto \psi(t)$.

We mention experimental evidence that a) cognition is not based on continuous-time processes (a moment in "cognitive time" correlates with $\Delta_{\text{cogn}} \approx 100$ ms of physical time); b) different *psychological functions* based on groups of neural networks performing specific cognitive tasks operate on different scales of physical time. In [173, 176] mental time was described mathematically by using p-adic hierarchic trees; see also [157–160] for applications of p-adic numbers in mathematical physics.

5.4 Quantum-like Consciousness

The brain is a huge information system that contains millions of patterns of neural activation. It could not "recognize" (or "feel") all those patterns at each instant of time t. Our fundamental hypothesis is that the brain is able to create the QL-representations of neural patterns. At each instant of time t, the brain creates the QL-representation of its mental context C based on two supplementary mental *self-observables* a and b. Here $a = (a_1, ..., a_n)$ and $b = (b_1, ..., b_n)$ can be very long vectors; each of them consists of nonsupplementary dichotomous observables. The *reference self-observables* can be chosen by the brain in different ways at different instances of time. Such a change of the reference observables is known in cognitive sciences as a *change of the representation*.

A mental context C in the $a|b$-representation is described by the mental wave function ψ_C. We can speculate that the brain has the ability to feel this mental field, a field of probability amplitudes.

In such a model the state of consciousness is represented by the mental wave function ψ_C. It is a projection of neuronal mental activity. The latter forms subconsciousness. We can say that one has the classical *subconsciousness* and the QL consciousness. We remark that this is a rather unusual viewpoint. Typically the consciousness is considered as the classical part of the brain's functioning and subconsciousness as quantum.

The crucial point is that in my model the consciousness is created through neglecting an essential volume of information contained in the subconsciousness. Of course, it is not just a random loss of information. Information is selected through the algorithm QLRA: a context C is projected onto ψ_C.

The (classical) mental state of subconsciousness evolves with time $C \rightarrow C(t)$. This dynamics induces dynamics of the mental wave function $\psi(t) = \psi_{C(t)}$ in complex Hilbert space, "mental Schrödinger dynamics."

Postulate QLR. *The brain is able to create the QL representation of mental contexts, $C \rightarrow \psi_C$, by using the algorithm (QLRA) based on the formula of total probability with the interference term.*

5.5 The Brain as a Quantum-like Computer

We can speculate that the ability of the brain to create the QL representation of mental contexts, see Postulate QLR, induces the functioning of the brain as a QL computer.

Postulate QLC. *The brain performs computation-thinking by using algorithms of quantum computing in the complex Hilbert space of mental QL states.*

We emphasize that in our approach the brain is not a quantum computer, but a QL computer. On the one hand, a QL computer works totally in accordance with the mathematical theory of quantum computations (so by using quantum algorithms). On the other hand, it is not based on superposition of individual mental states. The complex amplitude ψ_C representing a mental context C is a special probabilistic representation of information states of the huge neuronal ensemble. In particular, the brain is a *macroscopic* QL computer. Thus the QL parallelism (unlike conventional quantum parallelism) has a natural realistic base. This is real parallelism in the working of millions of neurons. The crucial point is the way in which this classical parallelism is projected onto dynamics of QL states. The QL brain is able to solve *NP*-problems. But there is nothing mysterious in this ability: an exponentially increasing number of operations is performed by involving an exponentially increasing number of neurons.

5.6 Evolution of Mental Wave Function

We restrict our considerations to trigonometric mental contexts (QL contexts producing the cos-interference). The mental wave function $\psi(t)$ evolves in complex Hilbert space $\mathcal{H}$ (space of probability amplitudes). The straightforward generalization of quantum mechanics implies the *linear* Schrödinger equation, see (2.43):

$$i\frac{d\psi(t)}{dt} = \hat{H}\psi(t), \quad \psi(0) = \psi_0, \tag{5.5}$$

where $\hat{H} : \mathcal{H} \to \mathcal{H}$ is a self-adjoint operator in the Hilbert space of mental QL states.

For example, let us consider a QL Hamiltonian, cf. (2.47):

$$\hat{H} \equiv H(\hat{a}, \hat{b}) = \frac{\hat{b}^2}{2} + V(\hat{a}), \tag{5.6}$$

where $V : X \to \mathbf{R}$ is a "mental potential" (e.g. a polynomial). We call $\hat{H}$ the operator of *mental energy,* cf. [161, 175, 53]. Here $\hat{a}, \hat{b}$ are two self-adjoint operators. We recall that in the Växjö model the operator representation can be constructed for any pair of supplementary observables, see Section 4.2, (4.18), (4.10).

Denote by ψ_j stationary mental QL states: $\hat{H}\psi_j = \mu_j\psi_j$. Then any mental QL state ψ can be represented as a superposition of stationary states

$$\psi = k_1\psi_1 + k_2\psi_2, \quad k_j \in \mathbf{C}, \quad |k_1|^2 + |k_2|^2 = 1. \tag{5.7}$$

One might speculate that the brain has the ability to feel the presence in the state $\psi \equiv \psi_C$ of superpositions (5.7) of stationary mental QL states. In such a case superposition would be an element of mental reality. However, it seems not to be the case. Suppose that ψ_1 corresponds to zero mental energy, $\mu_1 = 0$. For example, such a QL state can be interpreted as the state of depression. Let $\mu_2 >> 0$. For example, such a QL state can be interpreted as the state of excitement. My internal mental experience tells that I do not have a feeling of superposition of states of depression and high excitement. If I am not in one of those stationary states, then I am just in a new special mental QL state ψ and I have the feeling of this ψ (representing some mental context C, i.e., $\psi \equiv \psi_C$) and not superposition.[12] Thus it seems that the expansion (5.7) is just a purely mathematical feature of the model. Of course, the brain uses the possibility to select a basis, e.g., the eigenvectors of the operator of mental energy, and to perform self-measurements in this basis. However, as results of measurements, it will feel just these eigenfunctions and not their superposition ψ.

5.6.1 Structure of a Set of Mental States

In QM a state (wave function) ψ is represented by a vector belonging to the unit sphere S of a Hilbert space. In the two-dimensional case (corresponding to dichotomous observables, e.g., 'yes' or 'no' answers) the set of quantum states can be visualized by using the unit sphere in the three-dimensional real space $\mathbf{R}^3$, *Bloch's sphere*, see Section 4.3.

In our mental QL model, contexts producing trigonometric interference are represented by points in S. Suppose that there is given some set of cognitive contexts $\mathcal{P} \subset C^{tr}$, where the latter set consists of all trigonometric contexts[13] corresponding to the selected pair of two reference self-observables a and b. Let $S_{\mathcal{P}} = J^{b|a}(\mathcal{P})$, where $J^{b|a} : C^{tr} \to \mathcal{H}$ is the map corresponding to QLRA. Then the set of mental states is described by the $S_{\mathcal{P}}$. There is no reason to suppose that $S_{\mathcal{P}}$ coincides with the S. It is a fundamental problem to describe the set of QL metal states $S_{\mathcal{P}}$ for various classes of cognitive systems.

We might speculate that $S_{\mathcal{P}}$ depends essentially on classes of cognitive system. So $S_{\mathcal{P}}^{human}$ is not equal to $S_{\mathcal{P}}^{leon}$. We can even speculate that in the process of evolution

---

[12] We exclude abnormal behavior such as manic-depressive syndrome.

[13] Of course, the brain also could operate with non-trigonometric contexts, e.g., hyperbolic or even mixed hyper-trigonometric. We restrict modelling to trigonometric contexts to have a better analogy with conventional QM.

the set S_P has been increasing and S_P^{human} is the maximal set of mental states. It might even occur that S_P^{human} coincides with the Bloch sphere S.

5.6.2 Combining Neuronal Realism with Quantum-like Formalism

The main distinguishing feature of our QL approach to cognitive sciences is the possibility of combining neuronal realism with mathematical formalism of quantum mechanics (or its generalizations). In our model "quantum probabilistic waves" (represented in the mathematical model by complex probability amplitudes) are produced by ensembles of neurons. There is nothing mysterious in the wave-like dynamics of mental information. Such a dynamics (which we use to simulate the process of thinking) is the result of the ability of the brain to perform QL projection of the ocean of neuronal information. At each instant of (mental) time the brain selects two fundamental variables (selects a representation of the neuronal ocean[14]) and creates the image of activity of the neuronal ocean given by a complex probability amplitude (by applying QLRA producing a complex probability amplitude from the statistical data).[15] Our fundamental conjecture is that the brain operates (at least on the highest level of mental functioning) with such QL images by using algorithms of quantum computing. Thus one can call the brain a *QL computer.* Its functioning is mathematically described by the conventional theory of quantum computing, but physically it has nothing to do with the conventional quantum computer.

We can speculate that even collective cognitive systems (human societies, states, nations, groups of animals, birds, insects) are able to create QL probabilistic representations of information. One could say that such cognitive systems are driven by probabilistic QL waves. Finally, we remark that one could not exclude that such representations could be created by nonliving complex information systems. Our approach opens the way to *QL artificial intelligence.*

---

[14] Compare with *Solaris* by Stanislav Lem and especially with the corresponding film by Andrei Tarkovsky.

[15] Of course, it is assumed that the brain is able to collect this data. This collecting could not be performed instantaneously. Therefore we speak about moments of mental time which correspond to intervals of physical time.

Chapter 6
Experimental Tests of Quantum-like Behavior of the Mind

Recently Conte et al. [66–68] performed an experiment (proposed by the author of this book) suggesting that mental states may follow QL behavior. The conclusion of the experiment was that some kind of equivalence seems to exist between QL entities and corresponding cognitive entities. On this basis attempts can be made to use the mathematical theory of quantum mechanics (and its generalizations) to analyze the nature of cognitive entities. The aim of this chapter is to discuss some basic features of our previous experiment and to give evidence on the application of abstract quantum formalism to an analysis of cognition.

Another experiment to test QL probabilistic behavior in cognitive and social sciences was recently designed by Haven and me with the help of psychologists Rakow and Damjanovic (both at the Department of Psychology, University of Essex, UK) [207]. In this chapter we formulate the conjecture we want to test and we describe and discuss the proposed experiment.

Finally, in Section 6.6 we consider an interference-type experiment for the financial market. This experiment has not yet been performed. Moreover, it might need an essential modification to be adequate for real trade at the financial market.

6.1 Theoretical Foundations of Experiment

Some aspects of quantum mechanics deal with the link between human cognition and the physical world, so that indications of an ability of the theory to account for the relationships of mental states with external event were present in its early formulations. A correspondence on this subject took place in the years 1932–1958 between Pauli and Jung [155, 156].

Jung introduced the concept of *synchronicity* and Pauli devoted much consideration to it, concluding that *"it would be most satisfactory if physics and psyche could be seen as complementary aspects of the same reality"*. The concept of synchronicity states that a meaningful, although acausal, coincidence can occur of mental states with objective external events. In the same way Bohr was significantly influenced by the work of the psychologist James in the development of the principle of complementarity. He was well informed on some theses discussed by James in

A. Khrennikov, *Ubiquitous Quantum Structure*,
DOI 10.1007/978-3-642-05101-2_6, © Springer-Verlag Berlin Heidelberg 2010

his *Principles of Psychology*, and it was James's use of complementarity in psychology that possibly had a great influence on Bohr's subsequent formulation of this principle in quantum mechanics.

In this chapter I shall describe an experiment [66] designed in my work [180] to test the QL behavior of mind by using our approach based on the difference between classical and quantum formulas of total probability, see Section 5.2.2.

6.2 Gestalt Perception Theory

Let us explain the experiment in detail. It is well known that, starting in 1912, *Gestalt psychology* moved a devastating attack against the structuralism formulations of perception in psychology. The classical structuralism theory of perception was based on a reductionistic and mechanistic conception that was assumed to regulate the mechanism of perception. There exists for any perception a set of elementary defining features that are at the same time singly necessary and jointly sufficient in order to characterize perception also in cases of more complex conditions. The Gestalt approach introduced instead a holistic new approach, showing that the whole perception behavior of complex images can never be reduced to the simple identification and sum of elementary defining features defined in the framework of our experience.

During the 1920s and 1930s Gestalt psychology dominated in the study of perception. Its aim was to identify the natural units of perception, explaining it in a revised picture of the manner in which the nervous system works. Gestalt psychology's main contributions have provided some understanding of the elements of perception through the systematic investigation of some fascinating features, such as the causes of optical illusions, the manner in which the space around an object is involved in the perception of the object itself, and, finally the manner in which *ambiguity plays a role in the identification of the basic laws of the perception.* In particular, Gestalt psychology also made important contributions to the question of how to establish how it is that sometimes we see movements even though the object we are looking at is not really moving. As we know, when we look at something we never see just the thing we look at. We see it in relation to its surroundings (underlying context). An object is seen against its background. In each case we distinguish between the figure, the object or the shape, and the space surrounding it, which we call background or ground, see Fig. 6.1 and Fig. 6.2.

The psychologist Rubin was the first to systematically investigate this phenomenon, and he found that it was possible to identify any well-marked area of the visual field as the figure, leaving the rest as the ground.

However, there are cases in which the figure and the ground may fluctuate and one is forced to consider the dark part as the figure and the light part as the ground, and vice versa, alternately.

Only a probabilistic answer may be given for a selected set of subjects that will tend to respond on the basis of subjective and context-dependent factors. The

Fig. 6.1 (a) Ambiguity figure
1a. (b) Ambiguity figure 1b

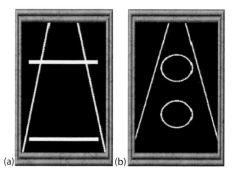

Fig. 6.2 Ambiguity figure 2

importance of the figure–ground relationship lies in the fact that this early work of Rubin represented the starting point from which Gestalt psychologists began to explain what today are known as the organizing principles of perception. A number of organizing or grouping principles emerged from such studies of ambiguous stimuli. Three identified principles may be expressed as similarity, closure and proximity. Gestalt psychologists attempted to extend their work also at a more physiological level, postulating the existence of a strong connection between the sphere of the experience and the physiology of the system, by admitting the well-known principle of isomorphism. This principle establishes that the subjective experience of a human being and the corresponding nervous event have substantially the same structure.

6.3 Gestalt-like Experiment for Quantum-like Behavior of the Mind

In the experiment, we examined subjects by Tests *a* and *b* in order to test QL behavior. For Tests *a* and *b* we used the ambiguity figures of Fig. 6.1, as they were widely employed in Gestalt studies:

a) Are these segments equal?
b) Are these circles equal?

Thus the a-test is based on the following cognitive task: look at Fig. 6.1a and reply to question (a). The b-test is based on Fig. 6.1b.

The reasons for using such ambiguity tests here to analyze QL behavior in perception may be summarized as it follows. First of all, the Gestalt approach was based on the fundamental acknowledgement of the importance of the context in the mechanism of perception. QL behavior formulates the same basic importance and role of the *context* in the evolution of the considered mechanism. Finally, we have seen that in ambiguity tests, the figure and the ground may fluctuate during the mechanism of perception. Thus, consequently, a non-deterministic (this is to say a QL) behavior should be involved.

Ninety-eight medical students of University of Bari (Italy) were enrolled in this study, with about equal distribution of females and males, aged between 19 and 22 years, after giving their informed consent to participate in the experiment. In the first experiment a group of 53 students was subjected in part to Test b (presentation of only Test b) and in part to Tests a and b (presentation of Test a and soon after presentation of Test b with prefixed time separation of about 2 s between the two tests). The same procedure was employed in the second and third experiments for groups of 24 and 21 students, respectively. All the students of each group were submitted to Test b or to Test a followed by Test b. The ambiguity figures of Test b or Test a followed by b appeared on a large screen for a time of only 3 s, and simultaneously the students were asked to mark on a previously prepared personal schedule their decision as to whether the figures were equal or not. Submission to students of Test a followed soon after by Test b had the objective of evaluating whether the perception of the first image (Test a) could alter the perception of the subsequent image (Test b). All the experiments were computer assisted and in each phase of the experimentation the following probabilities were calculated:

$$p^b(+), \quad p^b(-), \quad p^a(+), \quad p^a(-),$$
$$p(b = +|a = +), \quad p(b = -|a = +), \quad p(b = +|a = -), \quad p(b = -|a = -).$$

Here the role of context, say C, is played by the selection procedure of a sample for the experiment. All probabilities depend on C.

A statistical analysis of the results was performed in order to ascertain whether coefficients of supplementarity[1] λ_β are nonzero or zero in the case of our measurements of mental observables as they were performed by us with Tests b, a and $b|a$. The first experimentation gave the following results

---

[1] We recall the coefficient of supplementarity λ_β given by (4.3), see also (5.1).

$$\text{Test } b : p^b(+) = 0.6923; \ p^b(-) = 0.3077,$$
$$\text{Test } a : p^a(+) = 0.9259; \ p^a(-) = 0.0741,$$
$$\text{Test } b|a : p(b = +|a = +) = 0.68; \ p(b = -|a = +) = 0.32, \quad (6.1)$$
$$p(b = +|a = -) = 0.5; \ p(b = -|a = -) = 0.5.$$

The calculation of conditional probability gave the following result with regard to $p^b(+)$:

$$p^a(+)p(b = +|a = +) + p^a(-)p(b = +|a = -) = 0.6666. \quad (6.2)$$

The second experimentation gave the following results:

$$\text{Test } b : p^b(+) = 0.5714; \ p^b(-) = 0.4286,$$
$$\text{Test } a : p^a(+) = 1.0000; \ p^a(-) = 0.0000,$$
$$\text{Test } b|a : p(b = +|a = +) = 0.7000; \ p(b = -|a = +) = 0.3000, \quad (6.3)$$
$$p(b = +|a = -) = 1.0000; \ p(b = -|a = -) = 0.0000.$$

The calculation of the conditional probability gave the following result with regard to $p^b(+)$:

$$p^a(+)p(b = +|a = +) + p^a(-)p(b = +|a = -) = 0.7. \quad (6.4)$$

Finally, the third experimentation gave the following results:

$$\text{Test } b : p^b(+) = 0.4545; \ p^b(-) = 0.5455,$$
$$\text{Test } a : p^a(+) = 0.7000; \ p^a(-) = 0.3000,$$
$$\text{Test } b|a : p(b = +|a = +) = 0.4286; \ p(b = -|a = +) = 0.5714; \quad (6.5)$$
$$p(b = +|a = -) = 1.0000, \quad p(b = -|a = -) = 0.0000.$$

The calculation of the conditional probability with regard to $p^b(+)$ gave the following result:

$$p^a(+)p(b = +|a = +) + p^a(-)p(b = +|a = -) = 0.6000. \quad (6.6)$$

It is seen that the mean value $\pm$ SD of $p^b(+)$ resulted in $p^b(+) = 0.5727 \pm 0.1189$ with regard to Test b and calculated using (6.1), (6.3) and (6.5), while instead a mean value of 0.6556 ± 0.0509 resulted for $p^b(+)$ when calculated with regard to the Test $b|a$ and thus using (6.2), (6.4) and (6.6). The two obtained mean values are different and thus give evidence for the presence of QL behavior in measurements of cognitive mental states as they were performed by testing mental observables by Tests b, a and $b|a$. The use of Student's t-test demonstrated that we had no more than a 0.30 probability that the obtained differences between the two estimated values of $p^b(+)$ by Test b and by Test $b|a$ were produced by chance. Thus with probability 0.70 coefficients of supplementarity are nonzero and, hence, students behave in a QL way (with respect to observables based on ambiguous figures). We also found that these coefficients are bounded by 1, so behavior is *trigonometric*.

As the final step, we may proceed now to calculate $\cos\theta_\beta$ on the basis of the coefficient of supplementarity λ_β given by (4.3); see also (5.1). In the case of our experimentation we obtained $\cos\theta_+ = -0.2285, \theta_+ = 1.8013$ and $\cos\theta_- = 0.0438, \theta_- = 1.527$, which are quite satisfactory phase results in order to admit QL behavior for the investigated mental states.

On the basis of the illustrated results, we concluded that we had preliminary evidence of the existence of QL behavior in the dynamics of some mental states. Luckily, we were able to capture mental conditions of subjects in which the context influenced decision in an essential way. We had equivalence between QL entities and corresponding cognitive entities.

6.4 Analysis of Cognitive Entities

The performed experiment suggests a QL behavior of cognitive entities. A consequence could be that cognitive entities as well as quantum entities exhibit a highly contextual nature. As well as quantum entities being influenced by the usual physical act of measurement, cognitive entities are also influenced by the act of measurement. In the case of cognitive entities, the measurement is characterized by the cognitive interaction. According to the fact that the character of our knowledge is noncomplete, it follows that the cognitive entities must have states with noncomplete character.

In the present chapter, on the basis of a behavioral similarity between cognitive and QL entities, we have been able to make direct use of an abstract QL formalism applied to cognitive entities. Moreover, we have been able to carry out accurate QL development of the cognitive entity under consideration. The numerical results obtained on the basis of the previous experiment give us the opportunity to delineate basic features of cognitive entities not known in the past. Let us follow this application in more detail. We can introduce the complex QL amplitude, which represents the state of our cognitive entity expressed in relation to some selected mental observables. Let us admit that we selected the mental observable b, pertaining to a given cognitive entity. Let us admit also that b, as previously fixed, could assume only two possible values ($b = +, -$). Such a complex QL amplitude will be produced by QLRA. The Born rule holds

$$|\psi(\pm)|^2 = p^b(\pm). \tag{6.7}$$

The QLRA-produced complex QL amplitude will represent the state of our cognitive entity in relation to the considered mental observable b.

The experiment we have performed gives us the opportunity to express in detail a methodological indication of the way in which researchers could be able to apply such a new development in future experiments. We will briefly reconsider the case of the experiment we have performed, showing how to write QL complex amplitudes and thus to give a QL characterization of the state of the cognitive entity that was employed in the experiment. Let us consider in detail what was confirmed to exist in the case of our experimentation. As we indicated previously, we managed to calculate two different values for $\cos\theta(+)$ and $\cos\theta(-)$, whose meaning is now clear. In the case of our experimentations we obtained $\cos\theta_+ = -0.2285, \theta_+ = 1.8013$ and $\cos\theta_- = 0.0438, \theta_- = 1.527$, which are quite satisfactory phase results in order to admit QL behavior for the investigated cognitive entity. As a final step, we may now proceed by a detailed calculation of the QL model of the mental state of the cognitive entity as characterized during the experimentation.

By using the obtained data, we can write a mental wave function $\psi = \psi_C$ of the mental state C of the group of students who participated in the experiment – corresponding to a mental context denoted by the same symbol C. QLRA produces

$$\psi(\beta) = \sqrt{\mathbf{P}(a = +)\mathbf{P}(b = \beta|a = +)} + e^{i\theta(\beta)}\sqrt{\mathbf{P}(a = -)\mathbf{P}(b = \beta|a = -)}. \quad (6.8)$$

The ψ is a function from the range of values $\{+, -\}$ of the mental observable b to the field of complex numbers. Since b may assume only two values, such a function can be represented by two-dimensional vectors with complex coordinates. Our experimental data give

$$\psi(+) = \sqrt{0.8753 \times 0.6029} + e^{i\theta(+)}\sqrt{0.1247 \times 0.5} \approx 0.7193 + i0.2431 \quad (6.9)$$

and

$$\psi(-) = \sqrt{0.8753 \times 0.3971} + e^{i\theta(-)}\sqrt{0.1247 \times 0.5} \approx 0.5999 + i0.2494. \quad (6.10)$$

Thus, in conclusion, let us see how our development enables us to start from experimental results obtained in the course of a programmed experiment to proceed to writing explicitly the states of the involved cognitive entities as mathematical functions. They are now in so explicit a form that we are able to analyze and to test them under all the required experimental conditions.

Entering briefly the field of very technical notation, we may illustrate here that, by performing the same experiment for every group of people having some special mental state C (or in our terminology mental context), we can calculate the wave function ψ_C giving a QL representation of the C. Thus there exists the map J, which maps mental states into QL wave functions. Such a mapping provides a mathematical representation of mental functions. A mental state is too complex an object for its complete mathematical description to be available, but we can formalize mathematically some features of a mental state by using the QL representation. We remark that any QL representation induces a huge reduction of information. In particular, the wave function ψ_C calculated in our experiment gives a very rough representation of the mental state C. The ψ_C contains just information on the ability of the students who took part in the experiment to perceive contexts on the b- and a-pictures. On the space of complex functions we introduce the structure of a Hilbert space $\mathcal{H}$ with the aid of the scalar product $\langle \phi, \psi \rangle = \phi(+)\bar{\psi}(+) + \phi(-)\bar{\psi}(-)$. Thus J maps the set of mental states into the $\mathcal{H}$. The mental observable b can be represented by the multiplication operator in $\mathcal{H} : \hat{b}\psi(\beta) = \beta\psi(\beta); \beta = \pm$. The mean value of the mental observable b in the mental state C can be calculated by using the Hilbert space representation $E[b|C] = \langle \hat{b}\psi_C, \psi_C \rangle$. In our concrete experiment, using experimental data, classically we have $E[b|C] = p^b(+) - p^b(-) = 0.1454$. The same result gives our QL model

$$\langle \hat{b}\psi_C, \psi_C \rangle = |\sqrt{0.8753 \times 0.6029} + e^{i\theta(+)}\sqrt{0.1247 \times 0.5}|^2 \qquad (6.11)$$
$$- |\sqrt{0.8753 \times 0.3971} + e^{i\theta(-)}\sqrt{0.1247 \times 0.5}|^2 = 0.1454.$$

Thus we arrive at a complete quantitative representation of the basic features of the cognitive entity that was engaged in our experimentation, allowing us, in particular, to calculate mean values of mental observables. Such a way of applying an abstract QL formalism to cognition could indicate interesting perspectives in the studies of mental processes.

6.5 Description of Experiment on Image Recognition

The experiment considered in this section deals with image recognition [207]. The images in the experiment concern pictures (only faces) of well-known British people – be it royalty, film stars or any other individuals who are deemed to be well known to the experiment participants. Immediately, at least three comments are in order here. First, so as to make sure that we create no bias in proposing locally based pictures, we must make sure that we sample the experiment participants from a pool of students who have lived for a sizable period of time in the UK. There exists a sizable pool of such experiment participants at the University of Essex. Second, we need to remark that working only with students may create a bias too. In similar experiments recorded in the psychology literature, a sample is taken from a much wider age group. For instance, in the experiment on face recognition run by Collishaw and Hole [65], the age group of experiment participants sampled is 18–50 years. Third, we only focus on showing faces of well-known people to the experiment participants. Our choice of using faces stems from the literature in psychology, which says that faces are the most distinctive key to a person's identity [44]. The same literature [44] also investigates what types of information are derived from looking at the features of a face. We do not expand on this here. In our experiment two features of image recognition are being compared:

1) time of processing of the images;
2) the ability to recognize a nonmoving image I_0 by analyzing a deformation I of it.

We denote the time of processing of images by the variable t. The ability to recognize deformations of pictures is indicated by the variable r. The conjecture we want to test is

Conjecture: *t and r are complementary (or, using our terminology, supplementary).*

6.5.1 Preparation

We need to define the context (or, using quantum terminology, state) preparation. The experimental context (state) C is given by a sequence of images $I_1, \ldots, I_m$. The experiment participants form a group G and each participant looks at the images $I_1, \ldots, I_m$ for a sufficiently short time to "learn" the images. In Collishaw and Hole's [65] experiment on face recognition, the learning takes several seconds per face. Also, we may want to be explicit about the size of G. For statistical inference we need of course a group size that is sufficient. In Collishaw and Hole's [65] experiment the size was 140 experiment participants (68 women and 72 men). After learning, we split the group G randomly into two equal subgroups G_1 and G_2. We made use of a sequence of face images of well-known UK-based people (this sequence includes unknown people too). This sequence is now at the Department of Psychology at the University of Essex. The image sequence (in excess of 30 pictures) is currently loaded on a computer terminal that experiment participants will use to conduct the experiment.

6.5.2 First Experiment: Slight Deformations Versus Short Exposure Time

A first experiment is performed with experiment participants from the first group G_1. Experiment participants are seated in front of a computer terminal. Pictures are shown on the terminal which are slight deformations $I'_1, \ldots, I'_m$ of initial images $I_1, \ldots, I_m$. Those deformations can be obtained by various means:

  i) blurring of the images;
 ii) scrambling of the images;
iii) inversion of the images.

The psychology literature on the subject of face deformation is quite well established, see [65] .

Experiment participants in group G_1 are thus subjected to the images $I'_1, \ldots, I'_m$ and also a few other images that are unknown (i.e. which have no relation to $I_1, \ldots, I_m$).[2] Let us denote this set of all these images as I_{G_1}.

The time window, i.e. the time of exposure which experiment participants are allowed for viewing each of the slightly deformed pictures (including the unknown

---

[2] In principle, we could proceed without such additional images. We add them to make the process of recognition more complicated.

pictures), in this first experiment is set to be *very short*. The width of the time window, w is a parameter of the experiment. The task the experiment participants have to fulfill is to indicate whether they recognize the images in I_{G_1} as modifications of the initial images $I_1, \ldots, I_m$. One of two buttons on the computer terminal is to be pressed.

One button corresponds to the statement: "this image of I_{G_1} corresponds to I_j" and the other button corresponds to the statement: "this image of I_{G_1} does not correspond to any of $I_1, \ldots, I_m$." The images, part of the array of images loaned to us by the Department of Psychology at the University of Essex, are of standardized size. It is assumed that the experiment participants sit at a fixed distance from the computer terminal. Before the experiment participants start the experiment, they receive a practice set of pictures so they can get used to the terminal and to the expected responses (recognize – do not recognize) they are supposed to give.

Let ω be an experiment participant from group I_{G_1} performing this task. We set $t(\omega) = 1$ if ω was able to give the correct answers for $x\%$ of images in I_{G_1}. And $t(\omega) = 0$ in the opposite case. Alternatively, we can use a measure that seems to be used in psychology as an indicator of recognition accuracy. This is the so-called d' score [240] and there is a range of values for d' indicating that an experiment participant has done an almost error-less job in recognizing the pictures properly (low 3 value of d') and a so-called 'chance level performance' [65] p. 899 (0 value of d').

We now find probabilities $P(t = 1|C)$ and $P(t = 0|C)$ through counting numbers of experiment participants who gave answers $t = 1$ and $t = 0$, respectively. We denote respective subgroups of experiment participants by $G_1(t = 1)$ and $G_1(t = 0)$, respectively. The first group consists of experiment participants who have the ability to perform image recognition *very quickly* and the second consists of experiment participants who do not possess that ability. Note that the emphasis in the first experiment is on the time variable.

6.5.3 Second Experiment: Essential Deformations Versus Long Exposure Time

The second experiment is performed with experiment participants from the group G_2 as well as the subgroups $G_1(t = 1)$ and $G_1(t = 0)$ of the first group G_1. We create essential deformations $I''_1, \ldots, I''_m$ of initial images $I_1, \ldots, I_m$ (i.e. we increase the changes per face compared to the first experiment). As in the first experiment, images are again added which are unknown (i.e. which have no relation to $I_1, \ldots, I_m$). We denote the set of essentially deformed images (with the 'unknown' images) I_{G_2}. The task the experiment participants have to fulfill is identical to the task described in the first experiment, i.e. the experiment participants in G_2 have to indicate whether they recognize the images in I_{G_2} as modifications of the initial images $I_1, \ldots, I_m$. One of two buttons on the computer terminal is to be pressed. One button corresponds to the statement: "this image of I_{G_2} corresponds to I_j" and the other button corresponds to the statement: "this image of I_{G_2} does not correspond to any of $I_1, \ldots, I_m$". The time window in experiment 2 is long (i.e. longer than the 2.5–3 s of experiment 1). The width of the time window w is a parameter of the experiment.

Let ω be an experiment participant performing this task. We set $r(\omega) = 1$ if ω was able to give the correct answers for $x\%$ of images in the series. And $r(\omega) = 0$ in the opposite case. We now find probabilities $P(r = 1|C)$ and $P(r = 0|C)$ by counting numbers of experiment participants in the group who gave answers $r = 1$ and $r = 0$, respectively. We denote respective subgroups of experiment participants by $G_2(r = 1)$ and $G_2(r = 0)$, respectively. The first group consists of experiment participants who have the ability to proceed with image recognition very carefully and the second group consists of experiment participants without such an ability. Note that the emphasis in the second experiment is on the carefulness of recognizing the deformation of an image.

We also find probabilities $P(r = \beta|t = \alpha); \alpha, \beta = 0, 1$, by counting the number of people in the group $G_1(t = \alpha)$ who gave the answer $r = \beta$. After this we calculate the coefficient of supplementarity λ and if it is less than unity, we find the angle θ which gives us the measure of supplementarity of variables t and r. It would be very interesting if λ were larger than one! In this case we would find experimental evidence of new nonclassical probabilistic behavior (which is neither quantum nor classical). Finally, we pay particular attention to the possibility that the coefficient of supplementarity λ depends on the parameter w and the parameter of deformation.

Sub-additivity of probability in a psychology context has already been investigated, albeit not from a quantum mechanical point of view, see [30, 243]. We may come to similar conclusions through proposed studies of QL behavior of probabilities in psychology and cognitive sciences: toward nonadditive probability from QL interference of probabilities.

By using QLRA we can represent the experimental context C by a complex probability amplitude ψ_C – the mental wave function. One should be careful with interpretation of ψ_C. It should not be simply identified with its "material support". namely, the collection of images. The ψ_C also contains information about the ability of students of the group G to recognize images in different regimes. If the matrix of transition probabilities were double stochastic we would be able to construct the QL representation of variables t and r by self-adjoint operators $\hat{t}, \hat{r}$. We remark that t is the time-type variable! It is well known that in conventional quantum mechanics one cannot define the time operator. Of course, our QL time operator $\hat{t}$ is not the time-coordinate operator. This is the operator corresponding to some special temporal characteristics of the brain.

Conclusion: *The experiment proposed here aims to test complementarity (supplementarity) between the two defined variables t and r. We have three parameters to deal with in this experiment: the time width of exposure, the deformation and the parameter λ, which depends on the former two parameters. Especially, in the case λ > 1 we would have revealed, by this experiment, that there exists some nonclassical probabilistic behavior – which is neither classical nor quantum.*

6.6 Interference Effect at the Financial Market?

A detailed review of applications of quantum mathematics to finances can be found in Chapter 10.

The crash in 2008 demonstrated (once again) that the description of the financial market by current financial mathematics cannot be considered totally satisfactory. We recall that nowadays financial mathematics is heavily based on the use of random variables and stochastic processes, which are described by Kolmogorov's measure-theoretic model for probability ("classical probabilistic model"). I speculate that the present (2009) financial crisis is a sign (a kind of experiment to test the validity of classical probability theory for the financial market) that the use of this model in finances should be either totally rejected or at least completed. One of the best candidates for a new probabilistic financial model is quantum probability or its generalizations, that is, quantum-like (QL) models. Speculations that the financial market may be nonclassical have been present in the scientific literature for many years. Our aim is to move from the domain of speculation to rigorous statistical arguments in favor of probabilistic nonclassicality of the financial market. I design a corresponding statistical test that is based on violation of the formula of total probability (FTP). The latter is basic in classical probability and its violation would be a strong sign in favor of QL behavior of the market. We point out that the experimental test to check a possibility of violation of FTP of the financial market can be considered as adaptation to finances of the general statistical test proposed in Section 5.2.2. A version has already been tested in cognitive science, see Section 6.3. It was shown that FTP (and hence classical probability theory) is violated in some experiments on recognition of ambiguous pictures.

Our experiment may be criticized by dealers working at the real market. We cannot exclude such a possibility. However, our experiment opens the door toward design of similar, maybe more realistic, financial experiments. As a first step, one may try to perform our experiment with students.

6.6.1 Supplementary ("complementary") Stocks

We recall that two observables are supplementary if

$$\mathbf{P}(b = \beta | a = \alpha) > 0 \tag{6.12}$$

for all α and β. The latter condition has no direct relation to QM. It can be used in any domain of science. However, by using QLRA we can represent probabilities by complex amplitudes and observables a and b by self-adjoint operators $\widehat{a}$ and $\widehat{b}$. Then (6.12) is equivalent to noncommutativity of these operators.

We recall the meaning of this condition. It can be equivalently written as

$$\mathbf{P}(b = \beta | a = \alpha) \neq 1. \tag{6.13}$$

Thus it is impossible to determine a value $b = \beta$ by fixing the value $a = \alpha$. The b-variable has some features that cannot be determined on the basis of features of the a-variable. Thus b contains additional, or supplementary, information.Therefore we call observables (from any domain of science) *supplementary* if (6.12) holds. As was already pointed out, we may call them complementary as Bohr did in QM. But, unlike Bohr, we do not emphasize *mutual exclusivity of measurements.* In principle, supplementary observables, unlike complementary ones, can be measured simultaneously. However, supplementary observables are also mathematically represented by noncommuting operators. We recall that our aim is to show the adequacy of the mathematical apparatus of QM for the financial market. Thus, we need not borrow even quantum ideology and philosophy.

We will consider supplementary stocks. Formally, one can determine whether two stocks, say A and B, are supplementary by using the formal definition (6.12). However, to do this, we should perform an experiment for a large ensemble of dealers. If, finally, one observes that transition frequencies are close to zero, it will imply that this pair of stocks is not useful for a future interference experiment. Therefore it is much better to use financial intuition to determine whether two stocks can be assumed supplementary.

6.6.2 Experiment Design

1) Select two supplementary stocks, say A and B.

2) Select of an ensemble Ω of dealers who are used to working with these two stocks. Its size N should be large enough.

3) Select an interval δ giving the average time between two successive financial operations.[3]

4) Define two observables: for a dealer $\omega \in \Omega$, $a(\omega) = +1$ if he has bought a packet of A-stocks during the period δ and $a(\omega) = -1$ if he has not.[4] The b-observable is defined in the same way.

5) Starting with some initial instant of time, say t_0, wait until $t_0 + \delta$. Then count all dealers who have bought during this period some packet[5] of A-stocks, i.e., all elements $\omega \in \Omega$ such that $a = +1$. Denote this number by $n^a(+)$. We define frequency probabilities

---

[3] If during some period of time T (e.g., depending on frequency of operating, one day, a month, or a year), a dealer made k operations at the financial market, then δ is equal to average of T/k with respect to all dealers from the ensemble Ω selected for the experiment.

[4] Even if his A-bid was present at the market, but did not match asked prices for this stock, then $a = -1$ as if he did not submit any A-bid.

[5] In the experiment under consideration the size of packet does not play any role. However, the experiment can be designed in a more complicated way, by including the size of a packet. In this way nonsignificant bids can be excluded from the game.

$$p^a(+) = n^a(+)/N, \quad p^a(-) = 1 - p^a(+).$$

In the same way we find $n^b(+)$ – the number of dealers whose B-bids matched asks at the market (during the same period $[t_0, t_0+\delta]$) – and define frequency probabilities $p^b(\pm)$.

6) On the basis of the previous a-measurement select from Ω sub-ensembles of dealers Ω_+^a – those whose A-bids were realized during the period $[t_0, t_0 + \delta]$ – and Ω_-^a – those whose A-bids did not match any asked price for the A-stock or those who did not bid anything for this stock. Denote the numbers of elements in these ensembles by N_+^a and N_-^a, respectively.

7) Wait until $t_0 + 2\delta$ and after this count all dealers from Ω_+^a whose B-bids were realized during the period $[t_0 + \delta, t_0 + 2\delta]$. These are elements $\omega \in \Omega_+^a$ for whom $b(\omega) = +1$. Denote obtained number by $n(+|+)$. We define frequency probabilities

$$p_{+|+} = n(+|+)/N_+^a, \quad p_{-|+} = 1 - p_{+|+}.$$

They have the meaning of conditional probabilities. For example, $p_{+|+}$ is the probability that a randomly chosen dealer first bought a packet of the A-stocks and then a packet of the B-stocks.

In same way we define frequency probabilities

$$p_{+|-} = n(+|-)/N_-^a, \quad p_{-|-} = 1 - p_{+|-}$$

by making the b-measurement for dealers belonging to the sub-ensemble Ω_-^a.

8) Finally, define the interference coefficient

$$\lambda_\beta = \frac{p^b(\beta) - [p^a(+)p_{\beta|+} - p^a(-)p_{\beta|-}]}{2\sqrt{p^a(+)p_{\beta|+}p^a(-)p_{\beta|-}}}, \quad \text{where } \beta = \pm. \qquad (6.14)$$

It gives a measure of deviation from the classical formula of total probability.

9) An empirical situation with $\lambda \neq 0$ would yield evidence for QL behavior of the financial market: interaction of dealers and stocks. In this case, starting with the (experimentally calculated) coefficient of interference λ we can proceed either to the conventional Hilbert space formalism (if this coefficient is bounded by 1) or to so-called hyperbolic Hilbert space formalism (if this coefficient is larger than 1).

Chapter 7
Quantum-like Decision Making and Disjunction Effect

In this chapter we offer the QL representation of the probabilistic data obtained in two famous experiments in cognitive psychology: Shafir-Tversky [275] and Tversky-Shafir [295]. These experiments demonstrated violation of Savage's *Sure Thing Principle* (STP) [271]). This violation was called by Shafir and Tversky the *disjunction effect,* see also also Rapoport [266], Hofstadter [145, 146] and Croson [71].

7.1 Sure Thing Principle, Disjunction Effect

STP can be formulated like this:

STP *If you prefer prospect b_+ to prospect b_- if a possible future event A happens, and you prefer prospect b_+ still if future event A does not happen, then you should prefer prospect b_+ despite having no knowledge of whether or not event A will happen.*

Savage's illustration refers to a person deciding whether or not to buy a certain property shortly before a presidential election, the outcome of which could radically affect the property market. "Seeing that he would buy in either event, he decides that he should buy, even though he does not know which event will obtain", [271], p. 21.

The crucial point is that the decision maker is assumed to be *rational*. Thus the sure thing principle was used as one of foundations of *rational decision making* and rationality in general. It plays an important role in economics in the framework of Savage's *utility theory*.[1] It is well known that Savage's axiomatics implies the use of *subjective probability*. We recall that contextual probability in the Växjö model can be interpreted in various ways, e.g., as frequency probability or subjective probability, see Sect. 3.1.5. Probabilities in the formula of total probability (FTP) and its QL generalization, FTP with interference term (FTPQL), also can have different interpretations.

If one uses FTP or FTPQL for decision making, see Sections 7.8 and 7.9 for the latter, then probabilities (or at least a part of them) should be interpreted as

---

[1] But, it is not a postulate. It is derived from Savage's postulate P2, see [271].

subjective. However, if one uses these formulas to study experimental data, as we will do in Sect. 7.4, then probabilities are frequency or ensemble probabilities. In the latter case FTP is used as a test of probabilistic compatibility (PI), see Section 2.2. The main consequence of this test is that violation of FTP implies that, for decision making, cognitive systems (participating, e.g., in Shafir-Tversky [275] and Tversky-Shafir [295] experiments) use not the classical FTP, but FTPQL.

How does one apply classical FTP for decision making? Consider a Kolmogorovian probability space $\mathcal{P} = (\Omega, \mathcal{F}, \mathbf{P})$. Suppose there are given two mutually disjoint events A_+ and $A_- = \Omega \setminus A_+$. Decision function b is given by dichotomous (for simplicity) random variable $b : \Omega \mapsto \{+1, -1\}$. For example, somebody wants to predict probability $\mathbf{P}(b = +1)$ by assigning the subjective probability to the occurrence of A_+ (and hence A_-) and subjective probability to the decision $b = +1$ if A_+ occurs as well as the probability to the same decision if A_- occurs. Then by FTP

$$\mathbf{P}(b = +1) = \mathbf{P}(A_+)\mathbf{P}(b = +1|A_+) + \mathbf{P}(A_-)\mathbf{P}(b = +1|A_-). \qquad (7.1)$$

The crucial point of the use of FTP is that it is often easier to estimate probabilities under some special conditions, in our example, alternatives A_+ and $A_- = \Omega \setminus A_+$.

Sometimes, e.g., in experiment [295], one need not elaborate subjective probabilities for these alternatives, $\mathbf{P}(A_+)$ and $\mathbf{P}(A_-)$, see Section 7.4.2. They are given as objective, e.g., frequency or ensemble probabilities. In this situation only subjective probabilities under conditioning by these alternatives should be elaborated.

In other cases conditional probabilities can be found as frequency or ensemble probabilities; only (subjective) probabilities $\mathbf{P}(A_+)$ and $\mathbf{P}(A_-)$ should be assigned to alternatives.

Savage's STP is a simple consequence of FTP. Suppose now that, e.g., for $b = +1$, both conditional probabilities $\mathbf{P}(b = +1|A_+)$ and $\mathbf{P}(b = +1|A_-)$ are equal to 1. Then by (7.1)

$$\mathbf{P}(b = +1) = \mathbf{P}(b = +1|A_+ \cup A_-) = \mathbf{P}(A_+) + \mathbf{P}(A_-) = 1. \qquad (7.2)$$

Probability $\mathbf{P}(b = +1)$ can be considered as conditional probability with respect to disjunction of events A_+ and A_-, i.e.,

$$\mathbf{P}(b = +1|A_+ \cup A_-).$$

Experimentally observed [275, 295] violation of (7.2) was interpreted by Shafir and Tversky as a new probabilistic effect induced by creation of disjunction $A_+ \cup A_-$: the influence of disjunction (to the decision $b = +1$) cannot be reduced to separate influences of its counterparts. In cognitive psychology this situation is called the disjunction effect.

As was discovered by professor of cognitive psychology Jerome Busemeyer, for statistical data collected in Shafir-Tversky [275] and Tversky-Shafir [295] experiments, the classical FTP is violated. This was a great discovery! Busemeyer speculated that decision makers in these experiments (students) may use rules to make decisions that can be described by QM, see [48–50].

Inspired by the work of Busemeyer et al. [48], I applied the Växjö model to describe these experiments [208, 206, 213, 209]. On the basis of experimental statistical data I found the coefficients of interference. QLRA produced the corresponding ψ-functions – mental wave functions for groups of students participating in experiments.

My approach, which was called by Jerome Busemeyer the *"constructive wave function approach"*, demonstrated that the experimental statistical data can be produced by cognitive systems that use the wave function representation to make decisions.[2] Of course, it does not prove that students' brains really operate with probability amplitudes. Nevertheless, such a possibility of reconstructing an amplitude on the basis of results of decision making strongly supports *the hypothesis of QL representation of information in the brain*.

One important comment on the pioneer work by Busemeyer et al. [48– 51] should be made. Although the nonclassical character of data from Shafir-Tversky [275] and Tversky-Shafir [295] experiments was rightly pointed out, an attempt to apply directly conventional QM formalism was not totally justified. Matrices of $b|a$-contextual probabilities $\mathbf{P}^{b|a}$ corresponding to the mentioned experiments are *not doubly stochastic!* So, generalizations of QM should be used. Such generalizations appear as Hilbert space images of contextual probabilistic models, see Chapter 4 (the very end of the introduction and Remark 4.1) and a lot of details in [214].

We apply our contextual approach to describe mentioned experiments in terms of a variety of incompatible contexts that are involved, e.g., in the *prisoner's dilemma* (PD) or in more general games in which the disjunction effect can be found. We coupled this effect with violation of *the law of total probability*. It is evidently violated for the experimental data. Thus these data are nonclassical and the QL representation of these data can be useful. Moreover, we can find a numerical measure of contextual incompatibility or "nonclassicality" – the so-called coefficient of interference – as well as represent contexts that are involved in PD by probability amplitudes. It is done with the aid of QLRA, see Chapter 4.

Nowadays the use of quantum mechanics to study the disjunction effect and cognitive decision making is very popular (including applications to economics and the financial market). Besides the mentioned papers of Busemeyer and the coworkers [48– 51] and me [208, 206, 213, 209], there are important contributions by Danilov and Lambert-Mogiliansky [73–76] as well as Franco [109–114]. Finally, considerations of this chapter are generalized in [137].

In the following few sections the QL processing of information by cognitive systems is discussed, especially, QL decision making and both classical and QL rationality and ethics. These sections are of general theoretical and even philosophical character. See Sections 7.8 and 7.9 for a mathematical representation. The

---

[2] At the moment the hypothesis of the QL brain is not much supported by other studies in cognitive science and psychology. In any event we can create QL *artificial intelligence*.

reader can jump directly to Sect. 7.4.1 – the starting point of contextual analysis of Shafir-Tversky [275] and Tversky-Shafir [295] experiments.

7.2 Quantum-like Decision Making: General Discussion and Postulates

The algorithm QLRA represents contextual probabilities by wave functions (or normalized vectors of Hilbert space). We speculate that cognitive systems might develop (in the process of mental evolution) an ability to apply QLRA and to create QL representations of mental contexts. Thus, instead of operating with probabilities and analyzing (even unconsciously) probabilities of various alternatives, the brain works directly with mental wave functions (probabilistic amplitudes).

Such QL processing of information has the following advantages:

a) This is consistent processing of *incomplete information.*[3] The crucial point is that it is *consistent information cutoff.* Therefore, such processing does not induce "information chaos", especially under the assumption that all cognitive systems use the same QL representation.
b) This is linear (vector space) processing of information. From the purely mathematical viewpoint one can consider this procedure as *a linearization of a probabilistic representation of mental contexts.* In particular, the mental wave function evolves linearly. Such an evolution is described by the mental Schrödinger equation.[4]

We speculate that biological evolution induced the QL representation of information long before the creation of quantum mechanics by Planck, Einstein, Bohr, Heisenberg, de Broglie, Schrödinger, Dirac, and von Neumann.

If our hypothesis on QL processing of information by cognitive systems is correct, then it is natural to consider the QL process of decision making by cognitive systems, in particular, human beings. We recall that decision making is the cognitive process leading to the selection of a course of action among variations. Every decision-making process produces a final choice.

We describe briefly our model of decision making. Any decision is made within some mental context, say C, which is created in the brain on the basis of interactions with external conditions and self-interaction of neural processes in the brain. The brain represents a mental context by a mental wave function, probabilistic

---

[3] It should be recalled that in this book we do not debate such a fundamental problem of QM as its completeness. QM may or may not be complete. It is not important for us. We just apply QL mathematics to the description of processing of information by cognitive systems. In this case it is very natural to assume the existence of "hidden variables" on the neuronal level that are ignored in the QL representation of information by the brain.

[4] Thus, we surmise that the brain is able to linearize the mental world by the QL representation.

(complex or even hyperbolic) amplitude ψ_C. This mental wave function evolves linearly in the Hilbert state space: $\psi_C(t)$. The "decision-making operation" is represented by an observable, say b, taking values corresponding to different choices of action. This observable is represented by the brain as a linear operator (matrix) $\widehat{b}$. Probabilities of possible alternative decisions are produced by Born's rule for the mental wave function $\psi_C(T)$, where T is the instant of time when the decision is made

$$p_C^b(\beta) = |\langle \psi_C(T), e_\beta^b \rangle|^2,$$

where, as always, e_β^b is the eigenvector of $\widehat{b}$. Then the brain compares these probabilities for all $\beta \in X_b$ and it takes the decision $b = \beta(\max)$ corresponding to the maximal probability, see Sect. 7.8.

In such QL representation, the cognitive system selects a course of action among variations *purely automatically* (i.e., without applying the rule of reason – conventional Boolean logic) on the basis of a random generator reproducing the probability distribution of the QL observable b for the wave function $\psi_C(T)$. To calculate the probability that an observable b takes a fixed value, the brain should find the scalar product of the wave function $\psi_C(T)$ and the eigenvector corresponding to this value. Finally, the absolute value of the result of this procedure should be squared.

Postulates of QL Decision Making:

a) The brain applies QLRA to create the QL representation of mental contexts: a context C is mapped into its wave function ψ_C.
b) It generates dynamics of the mental wave function described by the Schrödinger equation.
c) It represents "decision-making observables" by linear operators.[5]
d) It applies Born's rule to find the probability distribution of a "decision-making observable".
e) Finally, the brain uses the classical decision-making scheme, see Sect. 7.8.

As has already been pointed out a few times, the QL representation is essentially more general than the quantum representation: matrices of "transition probabilities" need not be doubly stochastic. Moreover, some mental contexts might be represented not by complex probability amplitudes, but by hyperbolic (or even mixed hyper-trigonometric) amplitudes.

---

[5] Since a decision's spectrum consists of discrete alternatives, it is enough to operate in finite dimensional linear spaces, i.e., with matrices. In quantum mechanics observables are represented by self-adjoint operators, i.e., by symmetric matrices. However, we talk not only about the conventional quantum representation of cognitive entities, but about QL representation, which is based on the contextual approach. Contextual probabilistic setups could violate not only the classical probabilistic laws, for example, the law of total probability, but even the conventional quantum laws. For example, it might so happen that a mental observable could not be represented by a symmetric matrix.

7.2.1 Superposition of Choices

We remark that QL decision making also includes the QL dynamics of the mental state ψ_C. Of course, in the same way as in conventional quantum mechanics, by making a concrete choice among alternatives, a cognitive system disturbs the QL evolution, which is described, at least approximately, by Schrödinger's equation.

One could say that "collapse of the mental wave function" occurs at the instant of time $t = T$ when a decision is made. Unlike adherents of the conventional Copenhagen interpretation, I do not take collapse too seriously. In my model the ψ_C function is simply a special linear space representation of probabilistic data about the context C.

For example, let b take two values. These are two alternative decisions: $+1$, yes, or -1, no. Then the mental wave function and the decision maker determine two probabilities, p_+ and p_-. The concrete value, $b = +1$ or $b = -1$, is determined by these probabilities. For example, let b take the value $b = +1$. At this moment the Schrödinger evolution is interrupted. It starts again with a new mental wave function, which is equal to the eigenvector corresponding to the value $b = +1$. In accordance with quantum terminology we can say that during the period $0 \leq t \leq T$ the brain's mental state was in the superposition of two states $b = +1$ and $b = -1$.

Later we shall consider a more complicated process: a new context can be formed and represented by its own mental wave function. Evolution may start with it as an input, instead of with the eigenvector corresponding to the previous decision.

In general, a mental context C can be created not just for making the b-decision. Decision tasks can come later. Suppose that the brain has a collection of decision makers (self-observables) $a, b, ...$[6] The mental wave function $\psi_C(t)$ can be considered in the conventional quantum terminology as being in a superposition of all possible values for any observable. If the cognitive system should make the b-decision, then the b-superposition is reduced to a single value, e.g. $b = +1$. Suppose that operators (matrices) representing observables a and b do not commute. Then the eigenvector of b for the value $b = +1$ need not be at the same time an eigenvector for a. Hence, after taking the specific decision $b = +1$ the brain's state is still a superposition of all possible values for a.

Although we use the same terminology as in quantum mechanics, states superposition, its interpretation is totally different from the conventional one. Therefore, we prefer to speak about QL superposition of mental states and not quantum superposition. The first is described in purely classical terms (even Schrödinger's dynamics can be easily simulated by a classical neural network). Therefore it can be exhibited by *macroscopic systems*. The original quantum superposition is a "real superposition" of, e.g., states corresponding to two energy levels. It is not clear how it might be realized for macroscopic systems.

---

[6] It may be better to consider "activated decision makers". The total number of possible decision makers can be essentially larger. However, the majority of them are in the "sleeping mode."

7.2.2 Parallelism of Creation and Processing of Mental Wave function

It is clear that the brain cannot operate for long with the same context C. A series of Schrödinger's evolutions and "state updating" after a decision making can be interrupted when a new mental context C' is called forth by new external and internal signals. This context is represented by its own mental wave function $\psi_{C'}$, which evolves linearly in the Hilbert state space. The process of decision making and state updating is then repeated starting with $\psi_{C'}$.

If the brain's evolution is executed properly from the point of view of the information-processing architecture, then it is natural to assume that creation of a new context and its QL representation can go in parallel to processing, decision making and state updating based on the previous context C.

We consider two domains of the brain, classical and QL. In principle, each domain can be distributed through the brain. In the classical domain a probabilistic image of a mental context C is created, see Sect. 5.3.2. Then these contextual probabilities are represented by QLRA of the mental wave function. This mental wave function is processed in the QL domain: Schrödinger's evolution, measurement, updating, and so on. The classical domain does not "sleep" meanwhile. It works with a new context, say C'. Its amplitude representation will be transmitted to the QL domain later.

It is natural to suppose the existence of a control center, which plays the role of a conductor for activities of these two domains. In particular, it should control consistency of time scales for state preparation and decision making. On the one hand, the brain saves a lot of computational resources by working only in the QL domain. Here dynamics is linear – in contrast to essentially nonlinear dynamics in the classical domain of the brain. However, new signals change mental context and it should be updated in the classical domain.

7.2.3 Quantum-like Rationality

If one defines rational behavior on the basis of the law of total probability, then QL behavior can be positively irrational, see Sect. 7.3 on rational behavior, PD and so on. However, the only reason for such an interpretation is common application of the law of total probability in modern statistics and decision making. Under the assumption that cognitive systems make decisions by the QL decision-making procedure, violation of "Boolean rationality" does not look surprising. One should be much more surprised that modern science (including economics and finance) has been able to proceed so far on the basis of assumptions of classical "Boolean rationality."

Therefore, one should consider deviations from "Boolean rationality" not as evidence of irrational behavior, but as evidence that cognitive systems are QL rational.

We point out another source of QL rationality. In addition to the advantages of QL processing of incomplete information, see Sect. 7.2, let us mention the presence of *social pressure to proceed in the QL way*. If a society consists of QL thinking cognitive systems, then any individual should use the QL reasoning to proceed consistently with respect to other members of such a QL society. An individual who tries to use a more detailed description of mental contexts and who attempts to build a classical-like complete representation of contexts could make decisions that would be, in fact, "more rational" (from the point of view of complete information processing). However, such an individual might be rejected by the QL society.

7.2.4 Quantum-like Ethics

We remark that "nonconsequential reasoning" has been widely studied in cognitive psychology, e.g., Rapoport [266], Hofstadter [145, 146], Tversky and Shafir [295, 275], Croson [71]. However, from the QL point of view such reasoning is not nonconsequential at all. It is consequential, but consequences are obtained through QL processing of information.

For example, a preference for cooperative, ethical decisions in PD is consequential from the viewpoint of QL probability. Hence, human ethics is, in fact, a consequence of the QL representation of mental contexts. If we were involved in purely classical probabilistic reasoning (based on classical Bayesian analysis), we would not be able to demonstrate such a "nonconsequential behavior" as in PD. We would behave as "cognitive automata" (like creations of AI). The essence of human behavior is in the presence of the QL representation of probabilistic reality. Likewise, cooperation may arise simply from the fact that mental wave function produces (by Born's rule) larger probabilities for cooperative actions.

In the absence of a decision-making task, the mental wave function evolves according to Schrödinger's equation. The generator of evolution is represented by a special QL observable – "mental Hamiltonian" – describing a mental analog of energy.

We suppose that human beings have such mental Hamiltonians that produce "ethical wave functions", $\psi_C(T)$, starting with a large variety of ψ_C. Such an "ethical mental Hamiltonian" is formed still in childhood under the influence of the social environment. We cannot exclude that some elements of the "ethical mental Hamiltonian" are encoded in the genome.

7.3 Rational Behavior, Prisoner's Dilemma

In game theory, PD is a type of non-zero-sum game in which two players can cooperate or defect (i.e. betray the other player). In this game, as in all game theory, the only concern of each individual player (prisoner) is maximizing his/her own payoff, without any concern for the other player's payoff. In the classic form of this game, cooperating is strictly dominated by defecting, so that the only possible equilibrium for the game is for both players to defect. In simpler terms, no matter what the other player does, one player will always gain a greater payoff by playing defect. Since in any situation playing defect is more beneficial than cooperating, all rational players will play defect.

The classical PD is as follows: Two suspects, A and B, are arrested by the police. The police have insufficient evidence for a conviction, and, having separated the prisoners, visit each of them offering the same deal: if one testifies for the prosecution against the other and the other remains silent, the betrayer goes free and the silent accomplice receives the full 10-year sentence; if both stay silent, then both are sentenced to six months in jail; if both betray, each receives a two-year sentence.

Each prisoner must make the choice of whether to betray the other or to remain silent. However, neither prisoner knows for sure what choice the other prisoner will make. So this dilemma poses the question: How should the prisoners act? The dilemma arises when one assumes that both prisoners only care about minimizing their own jail terms. Each prisoner has two options: to cooperate with his accomplice and stay quiet, or to defect from their implied pact and betray his accomplice in return for a lighter sentence. The outcome of each choice depends on the choice of the accomplice, but each prisoner must choose without knowing what his accomplice has chosen to do. Reflecting on strategy, normally, it is crucial to predict what others will do. *This is not the case here.* If you knew the other prisoner would stay silent, your best move is to betray, as you then walk free instead of receiving the minor sentence. If you knew the other prisoner would betray, your best move is still to betray, as you receive a lesser sentence than by staying silent. Betraying is a dominant strategy. The other prisoner reasons similarly, and therefore also chooses to betray. Yet by both defecting they get a lower payoff than they would get by staying silent. So rational, self-interested play results in each prisoner being worse off than if they had stayed silent.

This is the *principle of rational behavior*, which is basic for rational choice theory – the dominant theoretical paradigm in microeconomics. It is also central in modern political science and is used by scholars in other disciplines such as sociology. However, Shafir and Tversky found that players frequently behave irrationally.

7.4 Contextual Analysis of Experiments with Disjunction Effect

7.4.1 Prisoner's Dilemma

Each contextual model is based on a collection of contexts and a collection of observables. The following mental contexts are involved in PD:

Context C, representing the situation when a player has no idea about the planned action of the other player, "uncertainty context."

Context C_+^A, representing the situation when the B-player supposes that A will stay silent ("cooperate"), and context C_-^A, when B supposes that A will betray ("compete"). Such a type of the PD experiment was performed by Croson [71]. Another version of the PD experiment (which is also realistic) was performed by

Shafir and Tversky [275]. In their experiments the B-player was informed about real actions of the A-player.[7]

We can also consider similar contexts $C_{\pm}^B$.

We define dichotomous observables a and b corresponding to *actions* of players A and $B : a = +$ if A chooses to cooperate and $a = -$ if A chooses to compete; b values are defined in the same way.

A priori the law of total probability might be violated for PD, since player B is not able to combine contexts. If those contexts were represented by subsets of the so-called space of "elementary events", as is done in classical probability theory, then player B would be able to consider the conjunction of the contexts C and, e.g., C_+^A and to operate in the context $C \wedge C_+^A$ (which would be represented by the set $C \cap C_+^A$). But the very situation of PD is such that one cannot expect contexts C and $C_{\pm}^A$ to be peacefully combined. If player B obtains information about the planned action of player A (or even if he just decides that A will play in a definite way, e.g., the context C_+^A will be realized), then the context C is simply destroyed. It could not be combined with C_+^A.

We can introduce the following contextual probabilities:

$p_C^b(\pm) \equiv P(b = \pm|C)$ – probabilities for actions of B under the complex of mental conditions C.

$p_{\pm|+} \equiv P(b = \pm|C_+^A)$ and $p_{\pm|-} \equiv P(b = \pm|C_-^A)$ – probabilities for actions of B under the complexes of mental conditions C_+^A and C_-^A, respectively.

$p^a(\pm) \equiv P(a = \pm|C)$ – prior probabilities that B assigns for actions of A under the complex of mental conditions C.

As we pointed out, there is no prior reason for FTP to be true in this situation, and experimental results of Shafir and Tversky [275] demonstrated that this equality could be indeed violated. As was remarked, Shafir and Tversky did not proceed in the mathematical framework, so they did not appeal to FTP. It was done significantly later by Busemeyer.

Shafir and Tversky [275] performed the following PD-type experiment. Participants were told that they would play a two-person PD against another participant. In fact, contrary to what they had been told, participants played against a preprogrammed strategy. All participants were told that they had been randomly assigned to a bonus group, which meant that they would occasionally be given information about the other player's already-chosen move before they had to choose their own. Throughout the experiment, each participant saw three versions of each PD: one in which the other player's move was unknown, one in which the other player had cooperated and one in which the other player had defected.

---

[7] It was a little bit more complicated, see coming description of their experiment.

In the Shafir–Tversky [275] PD experiment we have

$P(b = -|C) = 0.63$ and hence $P(b = +|C) = 0.37$;

$p_{-|-} = 0.97$, $p_{+|-} = 0.03$; $p_{-|+} = 0.84$, $p_{+|+} = 0.16$.

As usually in probability theory, it is convenient to introduce the matrix of transition probabilities

$$\mathbf{P}^{b|a} = \begin{pmatrix} 0.16 & 0.03 \\ 0.84 & 0.97 \end{pmatrix}.$$

We point out that this matrix is *stochastic* (as it should be). But it is clear that the *matrix obtained by Shafir and Tversky is not doubly stochastic!*

Croson [71] performed a similar experiment with one important difference. Unlike in the original Shafir–Tversky experiment, participants were not informed that they belong to a "bonus group" and that they would occasionally get information about actions of their co-players (in advance), but it was proposed – in a subset of games – to guess the actions of co-players. So, it was a game with elicited rather than controlled beliefs. In such a PD experiment Croson obtained the following data:

$P(b = -|C) = 0.225$ and hence $P(b = +|C) = 0.775$; so the cooperation rate was essentially higher than in the original Shafir–Tversky experiment.

$p_{-|-} = 0.68$, $p_{+|-} = 0.32$; $p_{-|+} = 0.17$, $p_{+|+} = 0.83$.

Hence, the matrix of transition probabilities is

$$\mathbf{P}^{b|a} = \begin{pmatrix} 0.83 & 0.32 \\ 0.17 & 0.68 \end{pmatrix}.$$

We see an essential deviation from the original Shafir–Tversky experiment. It is especially interesting that both experiments were based on the same payoff matrix:

$$\begin{pmatrix} 75, 75 & 25, 85 \\ 85, 25 & 30, 30 \end{pmatrix}.$$

Croson's experiment is very important in our mental contextual model. A mental context can be changed not only by a "real change of the world", but even by the brain's self-measurement. Even by imaging something the brain changes its state of mind, mental context.

In [71] an asymmetric version of PD was performed. Here the payoff matrix had the form

$$\begin{pmatrix} 85, 65 & 35, 75 \\ 95, 15 & 40, 20 \end{pmatrix}.$$

Here:

$P(b = -|C) = 0.375$ and hence $P(b = +|C) = 0.625$; so the cooperation rate was essentially higher than in the original Shafir–Tversky experiment.

$p_{-|-} = 0.65, \ p_{+|-} = 0.35; \ p_{-|+} = 0.47, \ p_{+|+} = 0.53.$

Hence, the matrix of $b|a$-contextual probabilities is

$$\mathbf{P}^{b|a} = \begin{pmatrix} 0.53 & 0.325 \\ 0.47 & 0.65 \end{pmatrix}.$$

7.4.2 Gambling Experiment

Tversky and Shafir [295] proposed to test the disjunction effect for the following gambling experiment. In this experiment, you are presented with two possible plays of a gamble that is equally likely to win 200 USD or lose 100 USD. You are instructed that the first play has completed, and now you are faced with the possibility of another play.

Here, a gambling device, e.g., roulette, plays the role of A; B is a real player, his actions are $b = +$, to play the second game, $b = -$, not to. The context C corresponds here to the situation when the result of the first play is unknown to B; the contexts $C_{\pm}^{A}$ correspond to the situations when B is informed of the results $a = \pm$ of the first play in the gamble.

Tversky–Shafir gambling experiment, version with the same group of students.

The data given here are from the experiment which was done for the *same group* of students, but under different contexts C_{+}^{A} (won-context), C_{-}^{A} (lost-context), C (uncertainty context). There was ten days pause between successive experiments. From Tversky and Shafir [295] we have

$P(b = +|C) = 0.36$ and hence $P(b = -|C) = 0.64$;

$p_{+|-} = 0.59, \ p_{-|-} = 0.41; \ p_{+|+} = 0.69, \ p_{-|+} = 0.31.$

We get the following matrix of transition probabilities

$$P = \begin{pmatrix} 0.69 & 0.59 \\ 0.31 & 0.41 \end{pmatrix}.$$

This matrix of transition probabilities is not doubly stochastic either, cf. with previously considered PD-type experiments.

Tversky–Shafir gambling experiment, between subject design.

In the same paper [295] Tversky and Shafir modified this gambling experiment. Three different populations, one for each context, were involved in the b-measurement. The data are

$P(b = +|C) = 0.38$ and hence $P(b = -|C) = 0.62$;

$p_{+|-} = 0.57, \ p_{-|-} = 0.43; \ p_{+|+} = 0.69, \ p_{-|+} = 0.31.$

We get the following matrix of transition probabilities:

$$P = \begin{pmatrix} 0.69 & 0.57 \\ 0.31 & 0.43 \end{pmatrix}.$$

7.4.3 Exam's Result and Hawaii Experiment

Tversky and Shafir considered the following psychological test demonstrating the disjunction effect. They showed that significantly more students report that they would purchase a nonrefundable Hawaii vacation if they knew that they had passed or failed an important exam than report they would purchase if they did not know the outcome of the exam.

The latter context is denoted by C and the "passed"-context by C_+^a and "failed"-context by C_-^a.

Here

$P(b = +|C) = 0.32$ and hence $P(b = |C) = 0.68$;

$p_{+|+} = 0.54, \; p_{+|-} = 0.57; \; p_{-|+} = 0.46, \; p_{-|-} = 0.43;$ and

$$\mathbf{P}^{b|a} = \begin{pmatrix} 0.54 & 0.57 \\ 0.46 & 0.43 \end{pmatrix}.$$

It is again not doubly stochastic.

7.5 Reason-Based Choice and Its Quantum-like Interpretation

Shafir and Tversky claimed that the disjunction effect is caused by the decision process of *reason-based* choice. Participants, instead of considering the consequences of their decisions, focus on reason to choose one thing versus another.

Go back to the example with Hawaii. If the exam were passed there would be a good reason to go to Hawaii – to celebrate. If the exam were failed, there would also be a good reason to go to Hawaii – to console oneself. However, before knowing the outcome of the exam, there is no reason to go to Hawaii; thus participants choose not to go. This dependence on reasons for choice leads to violation of STP.

In PD-type games, the information (real as well as obtained by imagination) on plans of the A-player induces reasons (for the B-player) that are absent in the absence of this information.

What does "reason-based choice" mean in the QL framework?

In our model the "state of mind" (representation of context C) is given by the vector $\psi \equiv \psi_C$ in Hilbert space. The process of thinking about the coming result of the exam can be interpreted as the process of measurement; we call the corresponding mental observable a. By obtaining a result $a = \alpha$, i.e., $a = +$, the brain finishes its self-measurement. It implies the creation a new context, namely, the selection context C_α, or (using the state formalism) *state reduction:* from ψ_C to the eigenvector $e_\alpha^a \equiv \psi_{C_\alpha}$ of the operator $\hat{a}$ representing the observable a, cf. Sect. 5.2.1. It is clear that statistics of results of measurements for observable b, $b = +$, to go to Hawaii, and $b = -$, not, is different for the state ψ and the state e_α^a.

Thus by looking for reasons the brain performs a self-measurement including the state reduction. The latter evidently changes statistics.

7.6 Coefficients of Interference and Quantum-like Representation

1a) **Tversky–Shafir gambling experiment,** Sect. 7.4.2; version with the same group of students. Here A-probabilities are equal: they were produced simply by a random generator imitating the first play of the gamble. Simple arithmetic calculations give

$$\delta_+ = -0.28, \quad \lambda_+ = -0.44.$$

The coefficient of interference is bounded by 1. Thus, the probabilistic phase $\theta_+ = 2.03$. We recall that $\delta_+ + \delta_- = 0$, see (3.16). So,

$$\delta_- = 0.28, \quad \lambda_- = 0.79.$$

This is again bounded by 1. Thus, the probabilistic phase $\theta_- = 0.66$. Since both coefficients of interference are bounded by 1, context C (uncertainty) is trigonometric, see Sect. 4.2: $C \in \mathcal{C}^{tr}$. Such behavior is typical for quantum systems. QLRA produces the complex probability amplitude – the mental wave function (in fact, two-dimensional vector)

$$\psi(+) \approx 0.59 + e^{2.03i}\, 0.54; \quad \psi(-) \approx 0.39 + e^{0.79i}\, 0.45. \tag{7.3}$$

1b) **Tversky–Shafir gambling experiment,** Sect. 7.4.2; between subject design. Here also A-probabilities are equal. Thus

$$\delta_+ = -0.25, \quad \lambda_+ = -0.4, \quad \theta_+ = 1.98;$$
$$\delta_- = 0.25, \quad \lambda_- = 0.69, \quad \theta_- = 0.81.$$

Thus (as one can expect) the uncertainty-context is again trigonometric. QLRA produces the complex vector with coordinates

$$\psi(+) \approx 0.59 + e^{1.98i}\, 0.53; \quad \psi(-) \approx 0.39 + e^{0.81i}\, 0.46. \tag{7.4}$$

As one can expect, the two complex vectors, (7.3) and (7.4), do not differ much.

2) **Shafir–Tversky PD experiment**, Sect. 7.4.1. In this PD experiment player B was given the information that player A had chosen to cooperate and to compete an equal number of times. Thus here A-probabilities are also equal. Here $\lambda_- = -0.31$ and hence the phase $\theta_- = 1.89$. However, $\lambda_+ = 3.98$. Thus interference is very high. It exceeds the possible range of the conventional trigonometric interference. This is *the case of hyperbolic interference!* Here the hyperbolic phase $\theta_+ = \text{arccosh} (3.98) = 2.06$.

This is the first example of hyperbolic interference! It shows that students are even more nonclassical than electrons and photons! It is one more sign that mental observables should be described by QL formalism and not by the conventional quantum one.

The B-brain (if it is really QL) represents the uncertainty context C in the PD game by the following hyper-trigonometric amplitude:

$$\psi(+) \approx 0.28 + e^{2.06j}\, 0.12; \quad \psi(-) \approx 0.65 + e^{1.89i}\, 0.7.$$

3) **Tversky–Shafir Hawaii experiment**, Sect. 7.4.3. Here the A-probabilities are equal as well. We have

$$\delta_+ = 0.17, \quad \lambda_+ = 0.3, \quad \theta_+ = 1.3;$$
$$\delta_- = -0.17, \quad \lambda_- = -0.37, \quad \theta_- = 2.$$

7.7 Non-double Stochasticity of Matrices of Transition Probabilities in Cognitive Psychology

We have seen that matrices of transition probabilities ($b|a$-contextual probabilities) constructed from experimental data of the Tversky–Shafir Game and Shafir–Tversky PD experiments are not doubly stochastic. The same is valid for the matrix obtained in the Bari experiment, Conte et al. [66]. On the other hand, matrices of transition probabilities that should be generated by conventional quantum mechanics in the two-dimensional Hilbert space are always doubly stochastic, see von Neumann [301], see also Sect. 2.3, Postulate 2.

We can present two possible explanations of this "non-doubly stochasticity paradox":

a) The statistics of these experiments are neither classical nor quantum (i.e., neither the Kolmogorov measure-theoretic model nor the conventional quantum model with self-adjoint operators could describe these statistics).

b) Observables corresponding to real and possible actions are not complete. From the viewpoint of quantum mechanics this means that they should be represented not in the two-dimensional (mental qubit) Hilbert space, but in Hilbert space of a higher dimension.[8]

Personally, I would choose explanation (a), and not merely because it is my own. It seems that actions of A and B in the PD could not be naturally split into sub-actions. Thus the "action-observables" could not be naturally represented by operators with degenerate spectra.

Of course, there are many brain-variables that are involved in PD decision making. However, the essence of creation of a QL representation is a selection of the most important variables. Other variables should not be included in the QL representation chosen for a certain problem.

Nevertheless, we cannot completely ignore the incompleteness conjecture of Busemeyer and Lambert-Mogiliansky. Here, we would immediately meet a really terrible problem: "How can we find the real dimension of the quantum (or QL) state space?" So, if this dimension is not determined by values of complementary observables a and b, then we should be able to find an answer to the question: "Which are the additional mental observables that would complete the model?" One should find complete families of observables $u_1^a, ..., u_m^a$ and $u_1^b, ..., u_m^b$, compatible with a and b, respectively.

We remark that in the case of the hyperbolic interference we would not be able to solve the "non-double stochasticity paradox" even by going to higher dimensions.

My conjecture (similar ideas were also proposed by Luigi Accardi and Dierk Aerts, at least in our conversations and our e-mail exchange) is that the laws of classical probability theory can be violated in cognitive sciences, psychology, social sciences and economics. However, nonclassical statistical data are not covered completely by the conventional quantum model.

My personal explanation is based on the evidence that violation of the formula of total probability does not mean that we should obtain precisely the formula of total probability with the interference term that is derived in the conventional quantum formalism.

Nevertheless, the conventional quantum formalism can be used as the simplest nonclassical model for mental and social modelling.

7.8 Decision Making

As we have seen, if for some context C, probability distributions for supplementary observables a and b are known, then the complex probability amplitude ψ_C representing C can be reconstructed by using QLRA. This was the problem of

---

[8] This latter possibility was pointed out to me by Jerome Busemeyer and Ariane Lambert-Mogiliansky during the workshop "Can quantum formalism be applied in psychology and economy?" (Int. Center for Math. Modeling in Physics and Cognitive Sciences, University of Växjö, Sweden; 17–18 September, 2007).

representation of probabilistic data by complex probability amplitude, see Sections
4.2 and 4.3. My conjecture is that the brain developed the ability for such a QL
representation of probabilistic data. In such a QL model the brain uses probability amplitudes for decision-making. We restrict our considerations to trigonometric
contexts, so probability distributions are complex-valued.

Consider the following situation. A (mental) context C is given. The brain makes
a decision about the b-attribute, given by, e.g., $b = \beta_1, \beta_2$, – so choosing between
$b = \beta_1$ and $b = \beta_2$. The crucial point is that it is assumed that another attribute, say
$a(= \alpha_1, \alpha_2)$, which is supplementary to b, is involved in the process of decision-
making. Since variables a and b are supplementary (under the context C), interfer-
ence angles $\theta = (\theta_{\beta_1}, \theta_{\beta_2})$ should be considered, see (4.6). In the PD this a-attribute
is related to actions of another prisoner. In the Tversky–Shafir gambling experiment
it is simply the (classical) random generator producing wins and losses. The latter example shows that "quantumness" (qualitatively encoded by the interference
angles) is not a feature of a (in fact neither of b), but it appears via interrelation of
a, b and the context C. Our scheme of QL decision-making is based on the assumptions that there are given (created by the brain of the basis of previous experience)

a) $b|a$-contextual ("transition") probabilities $p_{\beta|\alpha}$;
b) the probability distribution of the a : $p^a_C(\alpha)$;
c) the probability distribution of the phase angles $\theta = (\theta_{\beta_1}, \theta_{\beta_2})$: $p_C(\theta)$.

Thus all these distributions are given a priori. One should not always identify
prior probabilities with "subjective probabilities." The previous frequency experience plays an important role in determination of these probability distributions.

The brain uses the formula of total probability with the interference term to find
the b-probabilities. Under the assumption that the interference angle is θ_β, it produces the probabilities

$$p^b_C(\beta|\theta) = \sum_\alpha p^a_C(\alpha)p_{\beta|\alpha} + 2\cos\theta_\beta \sqrt{p^a_C(\alpha_1)p_{\beta|\alpha_1} p^a_C(\alpha_2)p_{\beta|\alpha_2}}. \qquad (7.5)$$

The crucial point of the decision-making scheme is their interpretation:

For each β, $p^b_C(\beta|\theta)$ is the probability that under the condition that the $b|a$-
interference angle is θ_β (for the context C) the decision $b = \beta$ is "right", i.e., it
would produce some form of reward.

By the (classical) Bayes' formula the brain finds the joint probability distribution

$$p_C(\beta, \theta) = p_C(\theta)\left(\sum_\alpha p^a_C(\alpha)p_{\beta|\alpha} + 2\cos\theta_\beta \sqrt{p^a_C(\alpha_1)p_{\beta|\alpha_1} p^a_C(\alpha_2)p_{\beta|\alpha_2}} \right) \qquad (7.6)$$

and finally the total b-probabilities

$$\bar{p}^b_C(\beta) = \int d\theta\ p_C(\theta)\ p^b_C(\beta|\theta). \qquad (7.7)$$

As the extension of the interpretation of conditional probabilities, the probability $\bar{p}_C^b(\beta)$ is considered as the probability that the decision $b = \beta$ is right.

In the present decision-making scheme the brain makes the $b = \beta_1$-decision if $\bar{p}_C^b(\beta_1)$ is larger than $\bar{p}_C^b(\beta_2)$ and vice versa, cf. [148], p. 54. The qualitative meaning of "larger" is determined depending on the cognitive system and may be the context C.

We should also mention another QL decision-making scheme. Comparing the probabilities $\bar{p}_C^b(\beta_1)$ and $\bar{p}_C^b(\beta_2)$ is an additional act of mental processing. It needs special neuronal and time resources. The processing might be especially complicated when these probabilities do not differ essentially. In such a situation a QL cognitive system might choose the regime of "automatic probabilistic decision-making", namely, by just using a (classical) random generator producing decisions β_1 and β_2 with the probabilities $\bar{p}_C^b(\beta_1)$ and $\bar{p}_C^b(\beta_2)$

Remark 7.1 (Comparison with classical decision making) We remark that a cognitive system τ_{CL} that uses the classical probabilistic processing of information can apply the conventional formula of total probability to predict the b-probabilities on the basis of "transition probabilities" $p_{\beta|\alpha}$ and a-probabilities $p_C^a(\alpha)$. Thus one can consider the proposed QL scheme as simply introduction of an additional – interference – parameter θ and modification of the formula of total probability. The main source of such a modification of the conventional statistical considerations is the impossibility of combining the context C with the selection contexts C_{α_j} and hence to get the probabilities $P(b = \beta|CC_{\alpha_j})$. As we have seen, a QL cognitive system τ_{QL} cannot proceed in the same way. The formula of total probability with the interference term contains not only the transition probabilities and the a-probabilities, but also phases, and the latter are unknown. Thus even by choosing e.g. prior probabilities $p_C^a(\alpha)$ (under the condition that the transition probabilities were obtained from the previous frequency experience), the τ_{QL} could not predict b-probabilities. The distribution $p_C(\theta)$ of the interference parameter θ also should be created on the basis of the previous experience or chosen by subjective reasons.

By using QLRA the cognitive system τ_{QL} can construct for each $\theta = (\theta_{\beta_1}, \theta_{\beta_2})$ the complex probability amplitude $\psi_{C,\theta}(\beta)$. Then the b-probabilities can be represented by using Born's rule:

$$\bar{p}_C^b(\beta) = \int d\theta \ p_C(\theta) \ |\psi_{C,\theta}(\beta)|^2. \tag{7.8}$$

7.9 Bayesian Updating of Mental State Distribution

Thus by our model the brain of τ_{QL} proceeds by using a mixture of classical and quantum probabilities. The whole Bayesian scheme is purely classical, "quantumness" appears in (7.8) only via Born's rule. In fact, this rule is a consequence of (7.5).

However, as always, there arises the problem of the choice of prior probability distributions. Since the transition probabilities and the a-probabilities are present even in the classical Bayesian framework, only the phase distribution $p_C(\theta)$ makes a new (QL) contribution. A QL cognitive system τ_{QL} should itself learn to choose $p_C(\theta)$ on the basis of the previous experience of the $b|a$ decision-making under the context C. Such learning can be performed by the (conventional) Bayesian updating procedure.

By combining Bayes' and Born's formulas, we obtain

$$p_C(\theta|\beta) = \frac{p_C(\beta, \theta)}{\bar{p}_C^b(\beta)} = \frac{p_C(\theta)|\psi_{C,\theta}(\beta)|^2}{\int d\theta p_C(\theta)|\psi_{C,\theta}(\beta)|^2}. \tag{7.9}$$

By following the Bayesian scheme τ_{QL} would like to maximize the probability $p_C(\theta|\beta)$, i.e., to construct a map $m : X_b \to \Theta, m(\beta) = \theta_{\max}(\beta)$. Since the denominator in (7.9) does not depend on θ, this problem is reduced to maximization of the joint probability density $p_C(\beta, \theta)$.

Suppose now that under the context C the cognitive system τ_{QL} made the decision $b = \beta$ and this decision was successful (so τ_{QL} got some form of reward). Then the τ_{QL} would update the distribution $p_C(\theta)$ by maximizing $p_C(\beta, \theta)$. To simplify considerations and to extract the main QL factor, we assume that the transition probabilities as well as the a-probabilities are fixed. So, optimization is considered only with respect to the interference angles θ.

We recall, see (4.13), that in the case of the doubly stochastic matrix of transition probabilities $\theta_{\beta_1} = \theta_{\beta_2} + \pi$ and hence we can consider the one-dimensional phase parameter θ.

Example 7.1 (Discrete distribution of phases) Some context C is chosen. Suppose that the transition probabilities as well as the a-probabilities are equal to 1/2. Here the formula of total probability with the interference term gives

$$p_C(\beta_1|\theta) = \cos^2 \theta/2; \quad p_C(\beta_2, \theta) = \sin^2 \theta/2.$$

We remark that these probabilities coincide with polarization (or spin 1/2) probabilities obtained in QM, see e.g. [148]. It should be emphasized that this is really a simple coincidence of mathematical formulas. Unlike, e.g., Marley and Hornstein [238], we do not consider physical quantum systems. We now consider the simplest nontrivial case of the parametric set consisting of two points, e.g. $\Theta = \{\theta_1 = \pi/2, \theta_2 = \pi\}$. So, this cognitive system reduced (on the basis of some information) phases under the context C to two possible angles. Hence, $\bar{p}_C(\beta_1) = 1/2(\cos^2 \pi/4 + \cos^2 \pi/2) = 1/4$, $\bar{p}_C(\beta_2) = 1/2(\sin^2 \pi/4 + \sin^2 \pi/2) = 3/4$. Thus under the assumption that all phases in Θ are equally possible, this cognitive system τ_{QL} gets that $\bar{p}_C(\beta_2)$ is essentially larger than $\bar{p}_C(\beta_1)$. Hence, τ_{QL} makes the decision $b = \beta_2$. If the result of this decision was positive (i.e. some form of reward was obtained), τ_{QL} would like to update the state distribution. Since $p_C(\beta_2, \pi/2) = 1/4$ and $p_C(\beta_2, \pi/2) = 1/2$, the cognitive system will put (in future decision-making) more weight on $\theta_2 = \pi$, e.g. the updated distribution could be $p_C(\pi/2) = 1/3, p_C(\pi) = 2/3$.

Example 7.2 (Continuous distribution of phases) Suppose that all transition probabilities are equal. Let us consider the uniform distribution of phases on $\Theta =$
$[0, 2\pi)$: $dp_C(\theta) = 1/2\pi d\theta$. Here $p_C(\beta_1, \theta) = 1/2\pi \cos^2 \theta/2$; $p_C(\beta_2, \theta) =$
$1/2\pi \sin^2 \theta/2$. Hence, $\bar{p}_C(\beta_1) = \bar{p}_C(\beta_2) = 1/2$. Thus a definite decision cannot
be made.

Example 7.3 Suppose that all transition probabilities are equal. Let us consider
the uniform distribution of phases on $\Theta = [0, \pi/2)$: $dp_C(\theta) = 2/\pi d\theta$. Here
$p_C(\beta_1, \theta) = 2/\pi \cos^2 \theta/2$; $p_C(\beta_2, \theta) = 2/\pi \sin^2 \theta/2$. Hence, $\bar{p}_C(\beta_1) = 1/\pi +$
$1/2$, $\bar{p}_C(\beta_2) = 1/2 - 1/\pi$. Thus the $b = \beta_1$ decision is preferred. For this deci-
sion the maximum is approached for $\theta = 0$. Therefore this cognitive system would
update $p_C(\theta)$ by concentrating it at the point $\theta = 0$.

7.10 Mixed State Representation

We remark that the former Bayesian considerations can be mathematically represented by using mixed quantum states. Let us consider the density matrix

$$\rho_C \equiv \int_\Theta d\theta p(\theta) \, \rho_{C,\theta},$$

$$\rho_{C,\theta} \equiv \psi_{C,\theta} \otimes \psi_{C,\theta}.$$

We obtain the representation

$$\bar{p}_C^b(\beta) = \mathrm{Tr} \, \rho_C \, \widehat{\beta}, \tag{7.10}$$

where $\widehat{\beta}$ is the orthogonal projector corresponding to the eigenvalue $b = \beta$. Thus
the quantity

$$\frac{\bar{p}_C^b(\beta_1)}{\bar{p}_C^b(\beta_2)} = \frac{\mathrm{Tr} \, \rho_C \, \widehat{\beta}_1}{\mathrm{Tr} \, \rho_C \, \widehat{\beta}_2} \tag{7.11}$$

is used in QL decision-making.

7.11 Comparison with Standard Quantum Decision-Making Theory

In this section we compare our approach with standard quantum decision-making
theory, see e.g. [141, 148, 147, 238] (and references therein).

a) Interpretation. The crucial difference is that our formalism is not about really
   quantum physical systems, but about QL systems. Thus we do not need quantum

sources of randomness, e.g. electrons or photons, to perform our QL decision-making. Moreover, the essence of QL behavior is not consideration of a special class of systems, but of a special class of contexts or, to be more precise: interrelation between contexts and observables.

b) Scheme of decision-making. We consider a specific scheme (motivated by PD) involving *two supplementary ("incompatible") observables* a and b. Moreover, in general one of them, namely, a, is a generalized quantum observable, i.e., it cannot be represented by a self-adjoint operator (symmetric matrix).

c) Mathematics. We consider a specific parametrization of a prior quantum state, namely, by the interference angle θ.

d) Application. We apply our model to modeling of the brain's functioning as a macroscopic QL system or, to be more precise: a macroscopic system performing specific interconnections between contexts and observables (inducing nontrivial interference).

7.12 Bayes Risk

As usual in quantum decision-making, we consider Bayes risk corresponding to the deviation function $W_\theta(\beta)$, see [148], p. 46:

$$\mathcal{R}_C^b \equiv \int_\Theta dp(\theta) \sum_\beta W_\theta(\beta) p_C^b(\beta|\theta) = \int_\Theta dp(\theta) \sum_\beta W_\theta(\beta) |\psi_{C,\theta}(\beta)|^2 = \quad (7.12)$$

$$\int_\Theta dp(\theta) \sum_\beta W_\theta(\beta) \operatorname{Tr}\rho_{C,\theta} \widehat{\beta}.$$

Typically in quantum decision theory the problem of finding Bayes decision rule is considered, e.g. [148], p. 46–50. However, we are not interested in this problem, since the decision-making operator $\hat{b}$ is considered as given.[9]

In our model the brain is interested in minimizing Bayes risk for the fixed observable b by variation of the prior distribution of interference phases.

We come back to Example 7.1. Now we do not fix the distribution of phases on $\Theta = \{\theta_1 = \pi/2, \theta_2 = \pi\}$. Here $p = p(\theta_1)$ and $1 - p = p(\theta_2)$ are parameters of the model. Suppose that the deviation function is $W_{\theta_j}(\beta_i) = \delta_{ij}$. Thus Bayes risk is $\mathcal{R}_C^b = p \, p_C^b(\beta_1|\theta_1) + (1 - p) \, p_C^b(\beta_2|\theta_2) = p \cos^2 \theta_1/2 + (1 - p) \sin^2 \theta_2/2 = p/2 + (1 - p) = 1 - p/2$. Thus Bayes risk is minimal for $p = 1$. Hence, the brain would modify the prior (mixed) mental state into the (pure) mental state $\psi_{C,\pi/2}$.

---

[9] Of course, it could also be modified in the process of the brain's functioning, but we do not consider this problem in the present book.

7.13 Conclusion

Violation of the law of total probability can be used to explain the disjunction effect. The QL representation can be applied to describe this effect. The essence of our approach is the introduction of a numerical measure of disjunction, the so-called interference coefficient. In particular, we found the interference coefficients for statistical data from Shafir–Tversky [275], Tversky–Shafir [295] and Croson [71] experiments coupled to PD. We also represent contexts of these experiments by QL probability amplitudes, "mental wave functions." We found that, besides the conventional trigonometric interference, e.g., Tversky–Shafir [295], so-called hyperbolic interference can be exhibited in cognitive science – Shafir–Tversky [275]. We also found that matrices of transition probabilities for these experiments are not doubly stochastic, as was found by Conte et al. [66]. Thus the probabilistic structure of cognitive science is not simply nonclassical, it is fundamentally richer than the probabilistic structure of quantum mechanics. Cognitive systems exhibit more complex probabilistic behavior than electrons or photons! We developed a QL model of decision making. In this model agents use not FTP, but its QL generalization to obtain probabilities for possible decisions and estimate risks.

Chapter 8
Macroscopic Games and Quantum Logic

It was Niels Bohr who said that he believed that the discovery of quantum physics really is something more than the discovery of the laws of microphysics. He claimed that some aspects (mainly complementarity) of quantum mechanics can be manifested in other branches of science (biology etc.). However, in spite of the fact that quantum formalism proved to be the best description of physical processes with molecules, atoms, atomic nuclei and elementary particles, it is not used much for the description of other phenomena, cf., however, Chapters 5–7.

In this chapter we consider an interesting application of quantum formalism – the quantum description of some games for macroscopic players. Such games we call *QL games*.

This chapter is based on papers written by A. Grib, his student K. Starkov and me [124, 125]. This chapter is essentially based on quantum logic. A reader who is interested not in quantum logic, but just in QL games can move directly to Chapter 9, in which another class of QL games will be presented without coupling to quantum logic.

A natural candidate in the search for a QL macroscopic system without the use of Planck's constant is some kind of game. This game must be organized such that disjunction and conjunction defined by the rules of the game are not Boolean. This means that one can define the rules for conjunction $\wedge$ and disjunction $\vee$ experimentally: $C = A \vee B$ if every time A-true, there follows C-true, every time B-true, there follows C-true. $D = A \wedge B$ if every time D-true, there follows A-true, every time D-true, there follows B-true. Then one looks for other properties of such C, D in the game considered:

Is C true if and only if A-true or B-true or can it be that C is true also in other cases (there is not "only if")?

Is the rule of distributivity confirmed in our game for all A, B, C?

To conclude the general discussion of why games are the most natural examples of looking for application of quantum formalism one can remark that even such a special feature of quantum physics as noncommutativity of operators can be found in games. In games one deals with acts and acts very often depend on the order. A typical example is that everybody understands the difference arising if one changes the usual order, putting on first the shirt and then the suit...

A. Khrennikov, *Ubiquitous Quantum Structure*,
DOI 10.1007/978-3-642-05101-2_8, © Springer-Verlag Berlin Heidelberg 2010

The payoff matrix for the so-called *Wise Alice* game will be written in terms of operators in finite-dimensional Hilbert space for the spin-one-half system. The average profit of the Wise Alice game will be calculated as the expectation value of the payoff operator in the tensor product of Hilbert spaces for Alice and her partner Bob. *Nash equilibrium points will be found for different situations of the Wise Alice game.*

We give explicitly the rules of the game Wise Alice. This especially concerns the angles for projections of spin operators for Alice and Bob used for different cases of the game. Then we shall give another example of the QL game using the lattice for the spin-one system (massive vector meson in quantum physics). In contrast to our example of the spin-one-half Wise Alice game, here one can see the role of nondistributivity more explicitly, because it can be expressed in the payoff matrix structure.

As we said before, the peculiar feature of these QL games will be the necessity to use not the Kolmogorovian probability measure, but the probability amplitude. QL interferences of the alternatives lead to new rules of calculation of average profits. That is why when comparison of *the average profit in cases of classical and QL games is made, the profit occurs differently for these cases.*

So these QL games demonstrate the situations where the formalism of quantum physics is applied to macroscopic games. Our examples are totally different from what is now widely discussed in many papers in the name of quantum games [100]. All examples with quantum coins, quantum gamblers etc. use, in one way or another, microobjects described by quantum physics as hardware, while in our examples everything is totally macroscopic.

However, some results obtained in the cited quantum game activity can be applied to our examples. In our examples it is the strategies of Alice and Bob that are described by the quantum formalism. Alice, Bob and their acts are totally macroscopic, there is no need to use the Planck constant for them. It is the set of frequencies of their acts that is calculated by use of the quantum formalism with wave functions different for Alice and Bob. The optimal strategy for both participants is described by the Nash equilibrium and it is characterized by the special choice of wave functions for Alice and Bob giving the maximal profit for what he/she can get independently of the acts of the other partner.

An interesting feature of our QL games is that their description by the quantum formalism makes necessary the use of different probabilistic spaces when measuring observables represented by noncommuting operators. Events (acts) are mathematically described by projectors in Hilbert space. It is these projectors that form the nondistributive lattice (quantum logic) on which the quantum probability or the probability amplitude is defined. On commuting projectors one defines usual probability measure by the Born rule. Typical for quantum theory, interferences of alternatives arise. The quantum rule for the average profit takes into account these interferences. That is one reason for the difference of the results of calculation of this average following quantum rules and rules of the standard probability calculus using Bayesian conditional probabilities.

To conclude this introduction, we point to the recent publication of Aerts et al. [9]. There they proposed another (in some sense more advanced) model of a game for macroscopic players that produces quantum statistics.

8.1 Spin-One-Half Example of a Quantum-like Game

The game Wise Alice is an example of the well-known game in which each of the participants names one of some previously considered objects (e.g. paper, scissors, rock). In the case that the results differ, one of the players wins from the other some agreed sum of money. The rules of the game are the following.

1. The participants of our game A and B, call them Alice and Bob, have a rectangular box in which a ball is located. Bob puts his ball in one of the corners of the box, but does not tell his partner which corner. Alice must guess in which corner Bob put his ball.
2. Alice can ask Bob questions supposing a two-valued answer: "yes" or "no". Different from many of the usual games, the rules of this game are such that in the case of a "yes" answer Alice does not receive any money from Bob. In the opposite case she asks Bob to pay her some compensation. This feature is described by the structure of the payoff matrix.
3. Unlike in other such games [248] Bob has the possibility of moving the ball to any of the adjacent vertices of the rectangle after Alice asks her question.

   This additional condition decisively changes the behavior of Bob, making him become active under the influence of Alice's questions. Owing to the fact that negative answers are not profitable for him, he moves his ball to the "correct" adjacent vertex, if possible. So if Alice asks the question "Are you in vertex 1?" Bob answers "yes" not only if he is in 1 but also when he was in 2 or 4 because of the possibility of reacting to the question of Alice and moving his ball. However, if the ball of Bob was initially in vertex 3 he cannot escape the negative answer, whichever vertex he moves his ball to, and he fails, leaving the ball in the same corner.

   The same rule is valid for any vertex. One must pay attention that in this case Alice not only gets the profit, but also obtains exact information on the position of the ball: the honest answer of Bob immediately reveals this position.

   Alice knows about the manipulations of Bob. Therefore Bob's negative answers are valuable to Alice, as only from such answers can Alice unambiguously check exclusive positions of Bob. So different vertices are incompatible (exclusive) relative to negative answers of Bob. This leads to experimentally defined conjunction for Alice.
4. The game is repeated many times, each time Bob putting his ball in some corner and Alice guessing which corner.
5. Due to rule 3 Alice can draw the graph of the game with four vertices and lines connecting them, leading to the Hasse diagram of the spin-one-half system in

which two noncommuting spin projections are measured. The reason for the nondistributive lattice represented by the Hasse diagram arising for events in the Wise Alice game is due to the special property of disjunction recognized by Alice experimentally. She sees that

$$1 \vee 2 = 2 \vee 3 = 3 \vee 4 = 1 \vee 3 = 2 \vee 4 = I,$$

where I is "always true." Here 1,2, etc. are corresponding questions of Alice.

It is this unusual property of disjunction that makes Alice consult her friend, a quantum physicist, for recommendations concerning the frequencies of her questions about this or that vertex.

6. Concerning the frequencies of Alice's questions and Bob's putting the ball in this or that vertex, one must formulate a special rule interpreting the idea of preferences.

The characteristic of the Hasse diagram is that it can be represented by projectors in Hilbert space nonuniquely in many different ways, by making unitary transformations of one of the noncommuting operators.

So assume Alice does not know the exact form of the rectangle used by Bob. Its diagonals, which can be taken as length 1 in some units, can form different angles θ_B. Depending on the angle, Bob, thinking that Alice will ask him about corner 1, will place his ball more frequently in the vertex adjacent to vertex 1, that is closer to vertex 1. This can be considered a psychological parameter of the game.

This angle is fixed for the game. Alice, not knowing the angle θ_B, uses the hypothesis that it is some θ_A and her friend – the physicist – makes calculations of her average profit using this θ_A. The choice of Alice of θ_A means that she thinks that the preference of Bob's frequencies will correspond to the shortest lines defined by θ_A.[1]

So because of fixation of preferences for Bob's reactions, the real free choice for Bob is the choice between alternative vertices on the diagonals of the rectangle 1–3, 2–4.

Let the payoff matrix of Alice have the structure of a four-by-four matrix h_{ik} representing Alice's payoffs in each of 16 possible game situations, so that one has some positive numbers a,b,c,d as her payoffs in those situations when Bob cannot answer her questions affirmatively (Table 8.1). Our game is an antagonistic game, so the payoff matrix of Bob is the opposite to that of Alice: $(-h_{ik})$.

Table 8.1 The payoff matrix of Alice

A/B	1	2	3	4
1?	0	0	a	0
2?	0	0	0	b
3?	c	0	0	0
4?	0	d	0	0

---

[1] Thus Alice well knows Bob's psychology, but she does not know the value of the angle θ_B.

The main problem of game theory is to find so-called points of equilibrium or saddle points – game situations that are optimal for all players at once. The strategies forming the equilibrium situation are optimal in the sense that they provide to each participant the maximum of what he/she can get independently of the acts of the other partner. More-or-less rational behavior is possible if there are points of equilibrium defined by the structure of the payoff matrix.

A simple criterion for the existence of equilibrium points is known: the payoff matrix must have an element that is maximal in its column and at the same time minimal in its row. It is easy to see that our game does not have such an equilibrium point. Nonexistence of the saddle point follows from the strict inequality valid for our game

$$\max_j \min_k h_{jk} < \min_k \max_j h_{jk}.$$

In spite of the absence of a rational choice at each turn of the game, when the game is repeated many times some optimal lines of behavior can be found. To find them one must, following [300], look for the so-called mixed generalization of the game. In this generalized game the choice is made between mixed strategies, i.e. probability distributions of usual (they are called differently from mixed "pure" strategies) strategies.

As the criterion for the choice of optimal mixed strategies one takes the mathematical expectation value of the payoff, which shows how much one can win on average by repeating the game many times. In usual classical games these expectation values of the payoff are calculated by using the Kolmogorovian probability, and the optimal strategies for Alice and Bob are defined as such probability distributions on the sets of pure strategies $x^0 = (x_1^0, x_2^0, x_3^0, x_4^0)$ and $y^0 = (y_1^0, y_2^0, y_3^0, y_4^0)$ that for all distributions of x, y the von Neumann–Nash inequalities are valid:

$$H_A(x^0, y^0) \geq H_A(x, y^0),$$
$$H_B(x^0, y^0) \geq H_B(x^0, y),$$

where H_A, H_B -payoff functions of Alice and Bob are the expectation values of their wins

$$H_A(x, y) = \sum_{j,k=1}^{4} h_{jk} x_j y_k, \quad H_B(x, y) = -\sum_{j,k=1}^{4} h_{jk} x_j y_k. \quad (8.1)$$

The combination of strategies satisfying the von Neumann–Nash inequalities is called the situation of equilibrium in Nash's sense. The equilibrium is convenient for each player; deviation from it can make the profit smaller. In equilibrium situations the strategy of each player is optimal against the strategy of his (her) partner.

The calculation of averages for the Wise Alice game, in which Bob reacts to her questions, must be different from that in the classical game. Instead of the usual probability measure one must use the probability amplitude (the wave function) and calculate probabilities for different outcomes using Born's formula. The reason for this is the following. If one defines the proposition of Alice that Bob's ball is located in vertex number k (defined by our rule 3) as α_k, then for any j, k one has

$$\alpha_j \vee \alpha_k = 1, \tag{8.2}$$
$$\alpha_j \wedge \alpha_k = 0.$$

Pairs of propositions with the same parity (α_1, α_3), (α_2, α_4) are orthocomplemented. The distributivity law is broken. So, for any triple of different j, k, l one has the inequality

$$(\alpha_j \vee \alpha_k) \wedge \alpha_l \neq (\alpha_j \wedge \alpha_l) \vee (\alpha_k \wedge \alpha_l). \tag{8.3}$$

Really, the left side of the inequality is equal to α_l, while the right side is zero. So the logic of Alice happens to be a nondistributive orthocomplemented lattice described by the Hasse diagram in Fig. 8.1.

In the Hasse diagram lines going up intersect in disjunction, lines going down intersect in conjunction. The lattice described by our Hasse diagram is isomorphic to the ortholattice of subspaces of the Hilbert space of the quantum system with spin one-half and the observables S_x, S_θ. For our case it is sufficient to take the real (not complex) two-dimensional space. So one can draw on the plane two pairs of mutually orthogonal direct lines $\{a^1; a^3\}$, $\{a^2; a^4\}$ with the angle θ between them coinciding with some angle θ_A or θ_R due to rule 6 of the game (Fig. 8.2).

Following the well-known constructions of quantum mechanics, we take instead of the sets of pure strategies of Alice and Bob the pair of two-dimensional Hilbert spaces $\mathcal{H}_A$, $\mathcal{H}_B$. Use of Hilbert space permits us to realize the nondistributive logic of our players without difficulty. To the predicate α_k put into correspondence the orthogonal projector $\widehat{\alpha}_k$. The same is done for Bob. Then one writes the self-adjoint operator in the space $\mathcal{H}_A \otimes \mathcal{H}_B$, which is the observable of the payoff for Alice

$$\widehat{H}_A = \sum_{j,k=1}^{4} h_{jk} \widehat{\alpha}_j \otimes \widehat{\beta}_k, \tag{8.4}$$

Let Alice and Bob repeat their game with a ball many times and let us describe their behavior by normalized vectors $\phi \in \mathcal{H}_A$, $\psi \in \mathcal{H}_B$, so that knowing them one can calculate the average according to the standard rules of quantum mechanics as

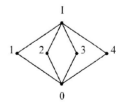

Fig. 8.1 The lattice of Alice's questions and Bob's answers

Fig. 8.2 Lattice of invariant subspaces of observer S_x, S_θ

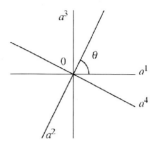

$$E_{\phi\otimes\psi}\,\widehat{H}_A = \sum_{j,k=1}^{4} h_{jk}\langle\widehat{\alpha}_j\phi|\phi\rangle\langle\widehat{\beta}_k\psi,\psi\rangle \tag{8.5}$$

$$= ap_1q_3 + cp_3q_1 + bp_2q_4 + dp_4q_2.$$

Here we take into account our payoff matrix. The operators $\widehat{\alpha}_1$, $\widehat{\alpha}_2$ can be written as two-by-two matrices

$$\widehat{\alpha}_1 = \begin{pmatrix} 1 & 0 \\ 0 & 0 \end{pmatrix}, \quad \widehat{\alpha}_2 = \begin{pmatrix} \cos^2\theta_A, & \sin\theta_A\cos\theta_A \\ \sin\theta_A\cos\theta_A & \sin^2\theta_A \end{pmatrix} \tag{8.6}$$

The projectors orthogonal to them are $\widehat{\alpha}_3$, $\widehat{\alpha}_4$, such that

$$\widehat{\alpha}_1 + \widehat{\alpha}_3 = \widehat{\alpha}_2 + \widehat{\alpha}_4 = I. \tag{8.7}$$

The wave function ϕ can be defined on the plane as some vector with the angle α, in general different from θ_A. Then the probabilities defined by Born's rule as projections on the corresponding basic vectors a_1, a_2, a_3, a_4 are

$$p_1 = \cos^2\alpha, \quad p_3 = \sin^2\alpha, \quad p_2 = \cos^2(\alpha - \theta_A), \quad p_4 = \sin^2(\alpha - \theta_A) \tag{8.8}$$

for Alice. For Bob one has some vector ψ with the angle β and θ_B for β_2 if β_1 has the same form as α_1:

$$q_1 = \cos^2\beta, \quad q_3 = \sin^2\beta, \quad q_2 = \cos^2(\beta - \theta_B), \quad q_4 = \sin^2(\beta - \theta_B). \tag{8.9}$$

The average profit is expressed as some function of α, β, dependent on parameters θ_A, θ_B:

$$F(\alpha, \beta) = a\cos^2\alpha\sin^2\beta + c\sin^2\alpha\cos^2\beta + b\cos^2(\alpha - \theta_A)\sin^2(\beta - \theta_B) \tag{8.10}$$
$$+ d\sin^2(\alpha - \theta_A)\cos^2(\beta - \theta_B).$$

This function is defined on the square $[0°, 180°] \times [0°, 180°]$. In paper [124] equilibrium points for different values of a, b, c, d and parameters θ_A, θ_B were found. These points correspond to Nash equilibria. Examples with two equilibrium points, one equilibrium point, and no such points at all were found. The results strongly depend on the difference of angle parameters and values of payoffs in the payoff matrix. Situations with two equilibrium points, as well as absence of such a point, are obtained for different values of angles and not equal a, b, c, d. If one takes $\theta_A = \theta_B = 45°$ and all a, b, c, d equal to 1 one has the simple solution $p_1 = 1, p_2 = 0.5, p_3 = 0, p_4 = 0.5$ for Alice, $q_1 = 1, q_2 = 0.5, q_3 = 0.5, q_4 = 0.5$ for Bob. So the wave functions for Alice and Bob in this case are just eigenfunctions of $\hat{\alpha}_1$ The payoff of the Wise Alice in this case is $E\ \hat{H}_A = 0.5$. Due to equivalence of all vertices one can take any other point as preferable for Alice and Bob, obtaining some other eigenstate of the spin projection operator. The preference here means that Bob never puts his ball into this vertex and Alice guesses this.The payoff of Alice in all these cases of eigenstates of spin operator projections will be the same. This can be compared with the result for the game called in [124] "the foolish Alice", who used standard probability calculus in the game with the same payoff matrix and the same rectangle being ignorant about Bob's reactions to her questions. Then Nash equilibrium corresponds to equal probability for any vertex and the result is $E\ H_A = 0.25$, which is smaller than that obtained by her "wise" copy. However, for the more general case [124] $\theta_A = 10°, \theta_B = 70°, a = 3, b = 3, c = 5$ one has $\alpha = 145.5°, \beta = 149.5°$ and one obtains for the Nash equilibrium in the "wise Alice" game $p_1 = 0.679, p_2 = 0.509, p_3 = 0.321, p_4 = 0.491, q_1 = 0.258, q_2 = 0.967, q_3 = 0.742, q_4 = 0.033$. Thus, differently from the more-or-less trivial case considered before, Bob and Alice here don't use the strategy of placing the ball in such a manner as to neglect totally one of the vertices. Their wave functions now are not just eigenstates of their spin operators.

8.2 Spin-One Quantum-like Game

Now consider the game described by the graph that coincides with the graph of the previous game but with the addition of one isolated point (Fig. 8.3).

This point is denoted as 0. The rules of the game generally are the same as in the previous one. Bob has the same possibility to react to Alice's questions by moving his ball to the adjacent vertex and only his negative answers are valid for Alice. If Bob is in the isolated point 0 he always gives her the "yes" answer, which is nonprofitable for Alice. A new rule will be added by us later making possible for Alice to ask two questions in some cases. We shall discuss it when writing the payoff matrix. Using the rule for drawing Hasse diagrams corresponding to the graph of automata one obtains Fig 8.4 for the case of the spin-one game.

Fig. 8.3 The graph of the spin-one game

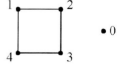

Fig. 8.4 Hasse diagram of
the spin-one game

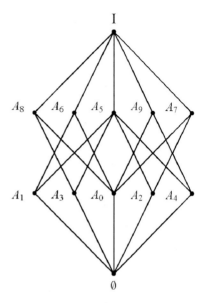

It is easy to see that elements A_1, A_3, A_2, A_4, A_5 form the same diagram as was considered for the spin-one-half game with the change of I to A_5. But addition of the point 0 leads not only to the appearance of the new logical atom A_0 in our lattice, but also to the appearance of the new level composed of A_5, A_6, A_7, A_8, A_9, which have the meaning of disjunctions

$$A_5 = A_1 \vee A_3 \vee A_2 \vee A_4, \quad A_9 = A_4 \vee A_0, \quad A_7 = A_2 \vee A_0.$$

Elements of the lattice can be represented by projectors on subspaces of $\mathbf{R}^3$ (Fig. 8.5).

To atoms correspond projectors on lines. A_1, A_2, A_3, A_4 are vectors in the plane. One has $A_0 \perp A_5$, $A_2 \perp A_4$, $A_1 \perp A_3$. The second level is represented by projectors on planes in $\mathbf{R}^3$: A_6 is the projector on the plane $A_0 A_1$, A_7 – on $A_0 A_2$, A_8 – on $A_0 A_3$, A_9 – on $A_0 A_4$. One has the orthogonality condition

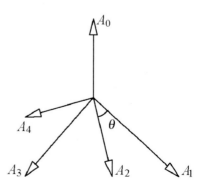

Fig. 8.5 Representation of
the lattice by subspaces in $\mathbf{R}^3$

$$A_6 \perp A_3, \quad A_8 \perp A_1, \quad A_7 \perp A_4, \quad A_9 \perp A_2.$$

Our lattice is the nondistributive modular orthocomplemented lattice. The nondistributivity is manifested due to

$$A_7 \wedge (A_1 \vee A_4) = A_7 \wedge A_5 = A_2 \neq (A_7 \wedge A_1) \vee (A_7 \wedge A_4) = \emptyset. \qquad (8.11)$$

This lattice can be represented by self-conjugate operators describing the spin-one system (massive vector meson) for which two noncommuting spin projections $\widehat{S}_z, \widehat{S}_\theta$ are measured. One has eigenvalues $S = 1, 0, -1$. The projector on the zero eigenvalue eigenvector is the same in $\widehat{S}_z, \widehat{S}_\theta$, that is why one has 5 atomic elements in the lattice. All projectors can be written as 3×3 matrices (see Table 8.2).

These are the well-known quantum mechanics operators of spin for the spin-1 system. The operators of observables of Bob are defined analogously, only the angle θ_A can be changed to θ_B. Now let us discuss the payoff matrix. Besides the same rules as were introduced for the spin-one-half QL game we add a new rule, arising naturally from the appearance of a new level in comparison with the spin-one-half game. The rule: Alice can ask not only one question, but two questions if the result of the first question corresponds to the disjunction $\vee$. Owing to her second question she can unambiguously guess where Bob's ball is located. For example Alice asks Bob the question: "Are you in 3?" The answer "no" means he is in either 0 or 1. Then she can ask the question: "Are you in 0?" The answer "no" will mean that he is at 1. As we formulated before, only negative answers are valid for Alice. So she receives from Bob in the considered case of two negative answers the sum of money v_1. But if she fails she receives nothing! If Alice asks the question 0 and the answer is "no", then Bob is in $1 \vee 2 \vee 3 \vee 4$ and by guessing correctly this disjunction Alice receives some u_0 for any case. If Alice does not want to risk then receiving the answer "no" on her question: "Are you in 3?" she doesn't ask the second question and receives some u_3. The same rule is true for any question. To write the payoff

Table 8.2 Matrices

$$A_0 = \begin{pmatrix} 0 & 0 & 0 \\ 0 & 0 & 0 \\ 0 & 0 & 1 \end{pmatrix} \quad A_1 = \begin{pmatrix} 1 & 0 & 0 \\ 0 & 0 & 0 \\ 0 & 0 & 0 \end{pmatrix} \quad A_2 = \begin{pmatrix} \cos^2 \theta_A & \sin \theta_A \cos \theta_A & 0 \\ \sin \theta_A \cos \theta_A & \sin^2 \theta_A & 0 \\ 0 & 0 & 0 \end{pmatrix}$$

$$A_3 = \begin{pmatrix} 0 & 0 & 0 \\ 0 & 1 & 0 \\ 0 & 0 & 0 \end{pmatrix} \quad A_4 = \begin{pmatrix} \sin^2 \theta_A & -\sin \theta_A \cos \theta_A & 0 \\ -\sin \theta_A \cos \theta_A & \cos^2 \theta_A & 0 \\ 0 & 0 & 0 \end{pmatrix} \quad A_5 = \begin{pmatrix} 1 & 0 & 0 \\ 0 & 1 & 0 \\ 0 & 0 & 0 \end{pmatrix}$$

$$A_6 = \begin{pmatrix} 1 & 0 & 0 \\ 0 & 0 & 0 \\ 0 & 0 & 1 \end{pmatrix} \quad A_7 = \begin{pmatrix} \cos^2 \theta_A & \sin \theta_A \cos \theta_A & 0 \\ \sin \theta_A \cos \theta_A & \sin^2 \theta_A & 0 \\ 0 & 0 & 1 \end{pmatrix} \quad A_8 = \begin{pmatrix} 0 & 0 & 0 \\ 0 & 1 & 0 \\ 0 & 0 & 1 \end{pmatrix}$$

$$A_9 = \begin{pmatrix} \sin^2 \theta_A & -\sin \theta_A \cos \theta_A & 0 \\ -\sin \theta_A \cos \theta_A & \cos^2 \theta_A & 0 \\ 0 & 0 & 1 \end{pmatrix}$$

matrix put in the left corner the question that, in those cases when Alice risks it, plays the role of the second question. Putting

$$u_0 < v_1, v_2, v_3, v_4 \tag{8.12}$$
$$u_1 < v_0, v_3$$
$$u_2 < v_0, v_4$$
$$u_3 < v_0, v_1$$
$$u_4 < v_0, v_2$$

where all u_i, v_k are positive numbers one can define the payoff matrix (Table 8.3).

Let us explain again some rules defining the payoff matrix. For example, Alice asks the question: "Are you in 0?" The answer is "no" ! Then she asks the second question: "Are you in 3?" The answer "no " means that Bob is in 1 and he pays v_1. If Alice first asks question 1 and gets the "no" answer, then asks 3 and again gets "no" she knows that Bob is in 0 and she gets v_0. The same applies if she asks 2

Table 8.3 The payoff matrix

A/B		0	1	2	3	4
0	0	0	u_0	u_0	u_0	u_0
	1	0	0	0	v_3	0
	2	0	0	0	0	v_4
	3	0	v_1	0	0	0
	4	0	0	v_2	0	0
1		u_1	0	0	u_1	0
	0	0	0	0	v_3	0
	2	v_0	0	0	0	0
	3	v_0	0	0	0	0
	4	v_0	0	0	0	0
2		u_2	0	0	0	u_2
	0	0	0	0	0	v_4
	1	v_0	0	0	0	0
	3	v_0	0	0	0	0
	4	v_0	0	0	0	0
3		u_3	u_3	0	0	0
	0	0	v_1	0	0	0
	1	v_0	0	0	0	0
	2	v_0	0	0	0	0
	4	v_0	0	0	0	0
4		u_4	0	u_4	0	0
	0	0	0	v_2	0	0
	1	v_0	0	0	0	0
	2	v_0	0	0	0	0
	3	v_0	0	0	0	0
	5	v_0	0	0	0	0
	6	0	0	0	v_3	0
	7	0	0	0	0	v_4
	8	0	v_1	0	0	0
	9	0	0	v_2	0	0

or 4 and then 3. Alice's questions 5,6,7,8,9 correspond to questions about disjunctions. Negative answers to them make it possible to guess where Bob is by only this one question. For example answer "no" to 5 (one must look at the Hasse diagram Fig. 8.4) means that Bob is in 0 and she gets v_0. A negative answer to 6 means that he is in 3 and she gets v_3, etc. The nondistributivity of the lattice is manifested in that, for example, owing to (8.11), by asking: "Are you in 0?" and receiving a negative answer she concludes that Bob is in $A_5 = A_1 \vee A_4$ and then asking: "Are you in 4?" and receiving a negative answer she concludes that he is in A_2 and gets the profit v_2. However, A_5 is also equal to $A_2 \vee A_4$ because of nonuniqueness of the disjunction, which is why Alice can come to the same conclusion without breaking her mind by the non-Boolean nondistributive logic! The payoff operator for Alice can be written as

$$
\begin{aligned}
\widehat{H}_A = {}& u_0 \widehat{A}_5 \otimes (\widehat{B}_1 + \widehat{B}_2 + \widehat{B}_3 + \widehat{B}_4) + u_1 \widehat{A}_8 \otimes (\widehat{B}_0 + \widehat{B}_3) \qquad (8.13)\\
& + u_2 \widehat{A}_9 \otimes (\widehat{B}_0 + \widehat{B}_4) + u_3 \widehat{A}_6 \otimes (\widehat{B}_0 + \widehat{B}_1) + u_4 A_7 (\widehat{B}_0 + \widehat{B}_2)\\
& + v_0 \widehat{A}_0 \otimes \widehat{B}_0 + v_1 \widehat{A}_1 \otimes \widehat{B}_1 + v_2 \widehat{A}_2 \otimes B_2 + v_3 \widehat{A}_3 \otimes \widehat{B}_3\\
& + v_4 \widehat{A}_0 \otimes \widehat{B}_4 + v_1 (\widehat{A}_9 \widehat{A}_6| \otimes \widehat{B}_1 + v_2 (\widehat{A}_9 \widehat{A}_7) \otimes \widehat{B}_2\\
& + v_3 (\widehat{A}_5 \widehat{A}_8) \otimes \widehat{B}_3 + v_4 (\widehat{A}_5 \widehat{A}_9) \otimes \widehat{B}_4
\end{aligned}
$$

The strategies of Alice and Bob are defined as vectors with angles

$$
\begin{aligned}
\phi = (\cos \alpha_1, \cos \alpha_2, \cos \alpha_3), \quad \psi - (\cos \beta_1, \cos \beta_2, \cos \beta_3) \qquad (8.14)\\
\alpha_1, \alpha_2, \beta_1, \beta_2 \in [0, \pi], \quad \cos \alpha_3, \cos \beta_3 \geq 0,\\
\cos^2 \alpha_1 + \cos^2 \alpha_2 + \cos^2 \alpha_3 = 1, \quad \cos^2 \beta_1 + \cos^2 \beta_2 + \cos^2 \beta_3 = 1.
\end{aligned}
$$

The average profit of Alice is calculated as

$$
\begin{aligned}
E\widehat{H}_A = {}& <\phi| \otimes <\psi|\widehat{H}_A|\psi> \otimes |\phi> = u_0 p_5 (q_1 + q_2 + q_3 + q_4)\\
& + u_1 p_8 (q_0 + q_3) + u_2 p_9 (q_0 + q_4) + u_3 p_6 (q_0 + q_1) + u_4 p_7 (q_0 + q_2)\\
& + v_0 p_0 q_0 + v_1 p_1 q_1 + v_2 p_2 q_2 + v_3 p_3 q_3 + v_4 p_4 q_4, \qquad (8.15)
\end{aligned}
$$

where

$$
\begin{aligned}
p_i = {}& <A_i \phi|\phi>, \quad q_i = <B_i \psi|\psi>, \qquad (8.16)\\
& p_0 = \cos^2 \alpha_3, \ p_1 = \cos^2 \alpha_1,\\
p_2 = {}& \cos^2 \theta_A \cos^2 \alpha_1 + \sin \theta_A \cos \theta_A (\cos^2 \alpha_1 + \cos^2 \alpha_2) + \sin^2 \theta_A \cos^2 \alpha_2,\\
& p_3 = \cos^2 \alpha_2, \ p_4 = 1 - p_0 - p_2,\\
p_5 = {}& 1 - p_0, \ p_6 = 1 - p_3, \ p_7 = 1 - p_4, \ p_8 = 1 - p_1, \ p_9 = 1 - p_2,\\
& q_0 = \cos^2 \beta_3, \ q_1 = \cos^2 \beta_1
\end{aligned}
$$

$$q_2 = \cos^2 \theta_B \cos^2 \beta_1 + \sin \theta_B \cos \theta_B (\cos^2 \beta_1 + \cos^2 \beta_2) + \sin \theta_B \cos^2 \beta_2,$$

$$q_3 = \cos^2 \beta_2, \quad q_4 = 1 - q_0 - q_2,$$

$$q_5 = 1 - q_0, \quad q_6 = 1 - q_3, \quad q_7 = 1 - q_4, \quad q_8 = 1 - q_1, \quad q_9 = 1 - q_2.$$

For different choices of θ_A, θ_B and different u_i, v_k one can obtain different Nash equilibria with some fixed values of α_i, β_k .

8.3 Interference of Probability in Quantum-like Games

Here we shall discuss the following question: If Alice doesn't know that Bob has the facility to move his ball when asked by Alice, can she get an understanding of this facility by observing the frequencies of his putting the ball in this or that vertex? This puts us into analysis of the (von Mises) frequency probability and our basic idea of the context dependence of probabilities. Similar to the case of PD games, in general FTP is violated and its QL modification, FTPQL, arises. The probabilistic phase θ is defined on the basis of the coefficient of interference. The latter provides a measure of deviation of FTPQL from FTP.

For the case of QL games (spin-one-half and spin-one cases) this θ is equal to zero (the coefficients of interference are equal to ± 1) and A, C correspond to two spin projections having values $+1$ or -1.

As we discussed before for the general case of Nash equilibrium with some θ_A, θ_B and nonequal values of a, b, c, d for the spin-one-half game, the state of Bob is some superposition of vectors of the basis for $\widehat{S}_{xB}$

$$|\psi > = c_1|e_1 > + c_3|e_3 > \tag{8.17}$$

and the probability for definite e_1 is

$$|c_1|^2 = | < \psi|e_1 > |^2. \tag{8.18}$$

In the basis of $\widehat{S}_{\theta B}$ we have

$$|\psi > = c_2|e_2 > + c_4|e_4 > \tag{8.19}$$

and

$$|e_1 > = c_\theta|e_2 > + \tilde{c}_\theta|e_4 >, \tag{8.20}$$

so, taking into account different signs for our coefficients we get

$$< \psi|e_1 > |^2 = |c_2 c_\theta + c_4 \tilde{c}_\theta|^2 \tag{8.21}$$
$$= c_2^2 c_\theta^2 + c_4^2 \tilde{c}_\theta^2 \pm 2|c_2||c_\theta||c_4| \cdot |\tilde{c}_\theta|.$$

In our case of real space all the coefficients are real and expressed as trigonometric functions of corresponding angles.

This formula can be understood as the generalization of FTP and if Alice, by looking at frequencies, can recognize it in Bob's behavior she can understand the QL nature of his game.

In the game Wise Alice one has two different supplementary observables: measuring S_x, meaning Alice asking questions about vertices of the diagonal "1–3" or measuring S_θ, Alice asking questions about vertices of the diagonal "2–4". But it is important to notice that owing to the structure of our game – the possibility of Bob changing the position of the ball by reacting to the question of Alice – we could not select, for example, elements with the property "1–3" without disturbing the property "2–4". So if one could consider the ensemble of possible game situations for Bob before Alice put her questions and call this ensemble S_0, then considering questions "1–3" of Alice as some filtration leading to ensemble S_1, the ensemble S_1 due to Bob's reactions will not coincide with S_0, i.e., the subensemble of nondisturbed Bob's positions.

Let us analyze from this point of view of contextually dependent subensembles the simple situation of the spin-one-half game with the rectangle and all a, b, c, d equal to one in the payoff matrix with the Nash equilibrium given by $p_1 = 1$, $p_3 = 0$, $p_2 = 1/2$, $p_4 = 1/2$.

What does it mean? It means that the initial ensemble of Bob was such that he never put his ball in 3. As to vertices 2 or 4 he put his ball equally in either 2 or 4. Without Alice's questions changing the whole situation one could think of his frequencies as if $N_1 - 2N_2 = 2N_3 = 2m$, so that the frequencies for nonperturbed ensembles are

$$\omega_4 = \omega_2 = \frac{1}{4}, \quad \omega_1 = \frac{1}{2}, \quad \omega_3 = 0. \tag{8.22}$$

However if Alice makes a selection of subensembles by asking questions concerning the ends of diagonals "1–3", "2–4" of the rectangle, the whole picture will be changed due to Bob's reactions. She will never see the ball in 3 as it was in the initial ensemble, but all balls will be moved to 1 when question 1 is asked. This leads to the situation $p_1 = 1$, $p_3 = 0$.

If questions 2,4 are asked and 1,3 are considered as meaningless, because of always a "yes" answer, then all the balls will be equally distributed between 2 and 4! That is the meaning of $p_2 = p_4 = 1/2$.

So her friend, the quantum physicist, supposing this Nash equilibrium strategy of Bob and calculating the average profit for Alice, will advise her to ask with equal frequency questions 2 and 4. Asking question 1 will be unprofitable for Alice because Bob will never give her a negative answer. The average profit will be 0.5 as we said before. It is easy to see that our rule 6 concerning interpretation of the angle for spin projection in terms of preferences and corresponding frequencies is directly manifested in the situation of eigenstate Nash equilibrium. In the more general case context dependence will be manifested not so trivially, interference terms must be taken into account and recommendations of the quantum physicist for Alice will be more sophisticated.

In conclusion, one should remark that surely not for all kinds of games with reactions hidden from one of the partners does one necessarily come to the quantum formalism. It is only for special kind of graphs of games that one arrives at Hasse diagrams of the quantum logical nondistributive modular orthocomplemented lattice. If the graph is such that the lattice is nonmodular or not orthocomplemented one will not have the QL structure. It is necessary to describe stochasticity in such cases by something different either from Kolmogorovian probability or from the quantum probability amplitude.

In particular, Alice and Bob can choose QL strategies that cannot be represented by complex amplitudes, but by hyperbolic ones. In such a case we obtain hyperbolic QL games. Moreover, it may happen that e.g. Alice's state is given by the complex amplitude and Bob's state by the hyperbolic one. In such a case we obtain hyper-trigonometric QL games. Has nothing been done in this direction, e.g. to find Nash equilibria?

8.4 Wave Functions in Macroscopic Quantum-like Games

Of course, the wave functions of Alice and Bob are not given by God. Moreover, in contrast to really quantum games, we do not use real quantum systems as sources of wave functions. We discuss in more detail the structure of QL games and in particular generation of the wave functions of Alice and Bob.

One must divide the game into two parts:

1. The preparation part, the rules of which are similar to those given in Section 8.1.
2. Measurement of two or more noncommuting operators on the described system. This second part consists of two or more classical games, the strategies of which must be those chosen by the partners in the first part. This "must" means some following of the "tradition" chosen in the first part.

In standard quantum mechanics, as is known, the frequencies of the results for measurements of different noncommuting observables with a definite prepared wave function are predetermined and cannot be arbitrary.

From the point of view of axiomatic quantum theory as the theory of quantum logical lattices, part 2 of our games corresponds to taking distributive sublattices of the initial nondistributive lattice with values of frequencies (or classical probabilities) prescribed by the quantum probabilistic measure (the wave function).

Preparation of the wave functions of Alice and Bob means defining frequencies of definite exclusive positions of Bob's ball and Alice's questions. These frequencies, however, differ from classical games in having less freedom in their definition. For example, in our first game imitating the spin-one-half system with two noncommuting observables, Alice and Bob, owing to constraint of the frequencies by the wave function, can only define one angle freely.

Besides the examples considered in our previous papers, an example of a spin-one-half system with three noncommuting observables of spin projections is dealt with. For this case one can look for imitations of Heisenberg uncertainty relations for spin projections in the macroscopic quantum game.

Let us describe these rules more explicitly.

1. *The preparation stage*

Alice and Bob have two quadrangles, one for Alice, another for Bob. Bob puts the ball in a vertex of his rectangle. Alice has to guess exactly in which vertex Bob put his ball. She does this by asking questions: "Is the ball in vertex "a"?" However Bob gives the answer "yes" not only if the ball is in "a" but also if the ball is in vertices connected by one arch with "a". It is only if the ball is in the opposite vertex that he cannot move it and definitely answers "no". This means that only negative answers of Bob are nonambiguous for Alice. In stage 1 Alice fixes the number of nonambiguous answers of Bob and calculates some frequencies for opposite vertices:

$$\omega_{1,B} = \frac{N_1}{N_1 + N_3}, \quad \omega_{3,B} = \frac{N_3}{N_1 + N_3}, \tag{8.23}$$

$$\omega_{2,B} = \frac{N_2}{N_2 + N_4}, \quad \omega_{4,B} = \frac{N_4}{N_2 + N_4}. \tag{8.24}$$

Now, to make the game symmetric, the same is supposed for Alice. Alice puts her ball in some vertex of the quadrangle and Bob must exactly guess the vertex. Only negative answers are nonambiguous for Bob.

To each graph of Fig. 8.6 corresponds the nondistributive lattice of Fig 8.1.

Definite frequencies $\omega_{a,B}$ mean that the wave function of Bob's ball is given as ψ_B and the representation of the lattice (Fig. 8.1) is defined such that two pairs of orthogonal projectors $\widehat{p}_1, \widehat{p}_3, \widehat{p}_2, \widehat{p}_4$ are chosen. Then we have $\omega_{a,B} = \langle \psi_b | \widehat{p}_a | \psi_b \rangle$, as it must be for the quantum spin-1/2 system with two observables – spin projections $\widehat{S}_z$ and $\widehat{S}_\theta$ with θ some angle being measured. Definite frequencies $\omega_{a,A}$ define the wave function of Alice's ball and some observables $\widehat{S}_z$, $\widehat{S}_\theta$ for Alice.

We can compare the process of determination of the wave function with establishing the tradition of the game with Bob. At the first stage (preparation) Alice tested Bob's behavior and she found interesting psychological features of Bob. She encoded Bob's behavior (moving of balls) by probabilities-frequencies, which are in

Fig. 8.6 Alice's and Bob's squares

turn encoded by the complex probability amplitude. After such a preparation Alice will play with Bob without setting more time for testing his behavior, just by using her wave function.

For example, one can consider preparation for marriage. At that stage it is important to find probabilistic characteristics of Bob's behavior. However, if Alice decided to marry Bob (in spite of all the problems with his behavior, i.e. without any hope of being sure of all his answers), after such a decision she just behaves according to her wave function, which was created during the preparation stage.[2] We remark that if Alice were sure of all the answers of Bob, she would be fine with classical noncontextual probability, i.e. to describe Bob without appealing to a complex probability amplitude.

One can also consider the following example. At the preparation stage some system of civil laws is created (on the basis of rather unfair behavior of people). After this preparation stage this system of laws is just applied. Thus one is no longer interested in "folk psychology." One is interested only in application of the established system of laws. This system of laws is represented by the complex (or hyperbolic or hyper-trigonometric) probability amplitude – the "wave function of folk psychology."

Our model can also be applied to describe the evolution of a complex biological system composed of various species. In our model this process has two main stages: a) the elaboration of rules of QL-gambling between species; b) QL-gambling. These stages cycle. Essential changes in environment induce new rules and new games.

We remark that by considering only frequencies (8.23), (8.24) Alice can determine only a special class of wave functions, i.e. without phase dependence. To obtain all possible wave functions, she should also collect conditional frequencies and apply QLRA.

2. Measurement stage

Two classical games are considered. Bob puts his ball only in vertices on one diagonal in the first game, let it be 1, 3. Alice asks questions trying to guess the position of Bob's ball. However, the frequencies of Bob putting his ball in one of the vertices must be $\omega_{1,B}$, $\omega_{3,B}$, defined at the first stage. The frequencies of Alice's questions must be $\omega_{1,A}$, $\omega_{3,A}$. Money is paid to Alice at this stage and the amount is fixed by the payoff matrix. In the second game Bob puts the ball in vertices 2,4 with frequencies $\omega_{2,B}$, $\omega_{4,B}$ and Alice asks questions with frequencies $\omega_{2,A}$, $\omega_{4,A}$. The profits from the two games are added.

The result will be given by use of the expectation value of the sum of projectors multiplied by the elements of the payoff matrix for the tensor products of two wave functions. The quantum game so formulated is an irrational game, which makes its theory different from the usual game theory. The payoff matrix is known to the players from the beginning, but as we see at the measurement stage it doesn't motivate

---

[2] We cannot exclude "collapse" of this mental wave function as a consequence of future interaction with Bob or other people.

their behavior. However, it can motivate somehow their behavior at the first stage, when the wave functions and "observables" are defined. Nash equilibria for fixed angles for observables can be understood as some "patterns" in the random choice of two players.

8.5 Spin-One-Half Game with Three Observables

Here we consider a more complicated game imitating a particle with spin-one-half, for which three noncommuting observables $\widehat{S}_x$, $\widehat{S}_y$, $\widehat{S}_z$ are measured. This case is interesting because here one can imitate Heisenberg uncertainty relations for spin projections in the case of our quantum game. For simplicity we consider that the same observables are measured by Alice and Bob (no difference in angles between projections is supposed). The Hasse diagram for this case is given in Fig. 8.7.
Here orthogonal projectors are $1-4, 2-5, 3-6$, which correspond for the spin-1/2 case to $\widehat{S}_x = \pm 1/2, \widehat{S}_y = \pm 1/2, \widehat{S}_z = \pm 1/2$.
The rule is the same, Bob can move his ball by one step, depending on Alice's question. For example, he can move to 1 from 2, 6, 3, 5 but not from 4 etc. For Alice the answer "no" on 1 means "Bob is at 4", the answer "no" on 2 means he is at 5, etc. The same is supposed for Alice's ball and Bob's questions. Representation of atoms of the Hasse diagram (Fig. 8.7) by projections is

$$A_1 = \begin{pmatrix} 1 & 0 \\ 0 & 0 \end{pmatrix} \quad A_2 = \frac{1}{2}\begin{pmatrix} 1 & 1 \\ 1 & 1 \end{pmatrix} \quad A_3 = \frac{1}{2}\begin{pmatrix} 1 & -i \\ i & 1 \end{pmatrix}$$
$$A_4 = \begin{pmatrix} 0 & 0 \\ 0 & 1 \end{pmatrix} \quad A_5 = \frac{1}{2}\begin{pmatrix} 1 & -1 \\ -1 & 1 \end{pmatrix} \quad A_6 = \frac{1}{2}\begin{pmatrix} 1 & i \\ -i & 1 \end{pmatrix}$$

The graph of the game, showing the vertices in which Bob and Alice put their balls is given in Fig 8.8 and the payoff matrix in Table 8.4.
Then the payoff operator of the quantum game is

$$\widehat{P} = v_1 A_1 \otimes B_4 + v_2 A_2 \otimes B_5 + v_3 A_3 \otimes B_6 + v_4 A_4 \otimes B_1 + v_5 A_5 \otimes B_2 + v_6 A_6 \otimes B_3.$$

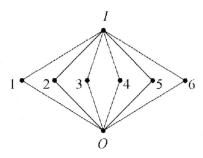

Fig. 8.7 The Hasse diagram of spin-one-half game with three observables

Fig. 8.8 The graph of "ball-vertices"

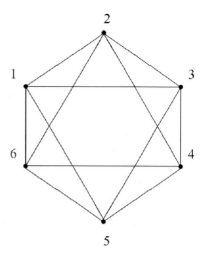

Table 8.4 The payoff matrix of Alice

$A\backslash B$	1	2	3	4	5	6
1	0	0	0	v_1	0	0
2	0	0	0	0	v_2	0
3	0	0	0	0	0	v_3
4	v_4	0	0	0	0	0
5	0	v_5	0	0	0	0
6	0	0	v_6	0	0	0

Here all B_i for Bob have the same form as A_i. The strategies of Alice asking questions and Bob putting the ball in the polygon (Fig. 8.8) are described as frequencies of choices in the "preparation part" given by wave functions, represented as vectors in complex Hilbert space: $\varphi_A = (\cos\alpha, e^{i\theta}\sin\alpha)$, $\psi_B = (\cos\beta, e^{i\omega}\sin\beta)$. So generally, different from real two-dimensional space in the previous example, one can take complex space in quantum spin-one-half physics. The average profit in three subsequent "measurement" games is

$$E_A = \langle\varphi_A| \otimes \langle\psi_B| \, \widehat{P} \, |\psi_B\rangle \otimes |\varphi_A\rangle.$$

This is calculated as

$$E_A = v_1 \cos^2\alpha \sin^2\beta + v_2 \frac{1+\cos\theta\sin 2\alpha}{2} \cdot \frac{1-\cos\omega\sin 2\beta}{2} +$$
$$+v_3 \frac{1+\sin\theta\sin 2\alpha}{2} \cdot \frac{1-\sin\omega\sin 2\beta}{2} + v_4 \sin^2\alpha \cos^2\beta +$$
$$+v_5 \frac{1-\cos\theta\sin 2\alpha}{2} \cdot \frac{1+\cos\omega\sin 2\beta}{2} + v_6 \frac{1-\sin\theta\sin 2\alpha}{2} \cdot \frac{1+\sin\omega\sin 2\beta}{2}.$$

So Nash equilibria can be found by analyzing the function $E_A(\alpha, \beta, \theta, \omega)$. The simplest case is when $\theta = \omega = 0$ and φ_A, ψ_B are real. For this case define

$$a = v_1, \quad b = v_4, \quad c = -\frac{v_2 + v_3 + v_5 + v_6}{4}, \quad d = \frac{v_2 + v_3 - v_5 - v_6}{4},$$

then $E_A = H(\alpha, \beta)$, where

$$H(\alpha, \beta) = a \cos^2 \alpha \sin^2 \beta + b \sin^2 \alpha \cos^2 \beta + c(1 - \sin 2\alpha \sin 2\beta) + d(\sin 2\alpha - \sin 2\beta).$$

To find Nash equilibria one must look for intersection points of curves of reaction of Bob and Alice. We investigated three cases.

- $a = 7, b = 1, c = -2, d = 1.5$. Nash equilibrium exists for $\alpha = \beta = \pi/8$. The value of the payoff at this point is equal to 2.
- $a = 1, b = 1, c = -2, d = 0$. No Nash equilibrium exists for this case.
- $a = 1, b = 10, c = -6, d = 4$. Nash equilibrium exists for $\alpha = 87.9°, \beta = 69.2°$, $E_A = 4.6$.

8.6 Heisenberg's Uncertainty Relations

As we said before, the game consists of two parts:

1. Preparation, when the nondistributive quantum logical lattice was used, leading to the choice of Alice and Bob of their wave functions.
2. Measurement, described by three different games, using orthogonal vertices of the graph (Fig. 8.7) and described by frequencies obtained from part 1.

As is well known from quantum mechanics, Heisenberg's uncertainty relations hold for spin projections, so that if

$$[\widehat{S}_x, \widehat{S}_y] = i\hbar \widehat{S}_z,$$

then for dispersions one has

$$D_\psi S_x \cdot D_\psi S_y \geqslant \frac{\hbar^2}{4}(E_\psi S_z)^2. \tag{8.25}$$

These relations for the graph are equivalent to the relation for frequencies obtained from the wave function:

$$p_1 p_4 p_2 p_5 \geq \frac{1}{16}(p_3 - p_6)^2. \tag{8.26}$$

Here the p_i-frequencies are for Alice. The same relation is valid for Bob. In our case

$$p_1 = \cos^2 \alpha, \quad p_2 = \frac{1 + \cos\theta \sin 2\alpha}{2}, \quad p_3 = \frac{1 + \sin\theta \sin 2\alpha}{2},$$

$$p_4 = \sin^2 \alpha, \quad p_5 = \frac{1 - \cos\theta \sin 2\alpha}{2}, \quad p_6 = \frac{1 - \sin\theta \sin 2\alpha}{2}. \tag{8.27}$$

Then (8.26) means $\sin^2 2\alpha \leq 1$, which is always valid.

In our case of three classical games with probabilities prescribed by (8.27) one can consider measuring three random variables A_1, A_2, A_3 taking values ± 1 and calculate dispersions and expectation values

$$D(A_1) = \sin^2 2\alpha, \qquad D(A_2) = 1 - (\cos\theta \sin 2\alpha)^2, \qquad E(A_3) = \sin\theta \sin 2\alpha.$$

So one obtains

$$D(A_1)D(A_2) \geq (E(A_3))^2 \tag{8.28}$$

equivalent to (8.26). Here differently from (8.25) we put $\hbar = 1$ and there is no $1/2$ as there was for the spin variable. However, if one includes in the notion of observable the payment defined by the payoff matrix, then for all equal v in the payoff matrix one can see that dimensional "price" can play the role of the Planck constant.

8.7 Cooperative Quantum-like Games, Entanglement

In all previous sections we have considered noncooperative games. Alice and Bob created their wave functions without cooperation. Therefore the total wave function was always the tensor product of two wave functions. However, at the stage of preparation Alice and Bob can cooperate. For example, both Alice and Bob want to become a couple, so they cooperate in testing their psychological behaviors. Such a preparation would not create a tensor product of two wave functions (at least for nontrivial cooperation). The wave function underlying the game would be an arbitrary element of the tensor product of state spaces of Alice and Bob,

$$\psi_{AB} \in \mathcal{H}_A \otimes \mathcal{H}_B.$$

The payoff is obtained as the average with respect to such a state ψ_{AB}. If this state is not factorizable one can talk about QL entanglement of strategies of Alice and Bob. Of course, such an entanglement has nothing to do with nonlocality. This is the result of simultaneous preparation.

Chapter 9
Contextual Approach to Quantum-like Macroscopic Games

In Chapter 8 the QL theory of macroscopic games was developed on the basis of quantum logic. These investigations were initiated by Andrey Grib and then continued in collaboration with me. They induced understanding that games for macroscopic players having all the distinguishing features of "really quantum games" (i.e., games that are based on microscopic sources of randomness such as pairs of entangled photons) can be easily constructed and represented by means of quantum logic. This QL (in fact, quantum logic) program on macroscopic games stimulated the author to apply the contextual statistical model to represent nonclassical (from the probabilistic point of view) games in the contextual form and then, using QLRA, map them into the complex (or even hyperbolic) Hilbert space. We shall do this in the present chapter. One of the consequences of considerations in this chapter is that in order to simulate QL games one does not need sources of quantum particles, e.g., photons. It is enough to use classical random generators.

9.1 Quantum Probability and Game Theory

We present a simple game between macroscopic players, say Alice and Bob (or in a more complex form - Alice, Bob and Cecilia), which can be represented in the QL way – by using a complex probability amplitude (game's "wave function") and non-commutative operators. The crucial point is that the games under consideration are so-called *extensive form games,* see e.g. [70]. Here the order of actions of players is important; such a game can be represented by a tree of actions. The QL probabilistic behavior of players is a consequence of *incomplete information,* which is available to e.g. Bob about the previous action of Alice. In general one cannot construct a classical probability space underlying a QL game – even for a QL game with two players. In a QL game with three players Bell's inequality for averages of payoffs can be written. It can be violated. The most natural probabilistic description in such a framework is given by our contextual probability theory. We shall use QLRA to find QL representations of extensive form games. The probabilistic structure of our game (for two players) can be considered as Gudder's probability manifold [126, 127] with the atlas having two charts.

A. Khrennikov, *Ubiquitous Quantum Structure,*
DOI 10.1007/978-3-642-05101-2_9, © Springer-Verlag Berlin Heidelberg 2010

The QL behavior can be produced by local gambling. However, QL gambling can be completed by interactions between players (of course, laws of special relativity are not violated).

9.2 Wine Testing Game

A restaurant has a good collection of (only) French and Italian wines of various sorts. Couples come to this restaurant for dinner, and to have more fun they play the following Wine Game, which consists of two wine tests.

A1) Alice selects a bottle (without telling her friend Bob the wine's name) and proposes that he tests the wine. A bottle of this wine is opened in the restaurant's kitchen; Bob gets just a glass of this wine. Alice asks him the question:

"Is it French or Italian?"

A2) If Bob answers (after testing) correctly, he gets some amount of money; if not, he loses money and Alice gets some amount of money.

The choice in A1 is not totally random, Alice has her own preferences (later she wants to share the chosen bottle with Bob).

In the second part of the game Alice and Bob interchange their roles, so Bob starts by choosing a bottle of French or Italian wine and so on.

We introduce for the first and second parts of the game the elements of the payment matrices

$$(h^b_{FF;k}, h^b_{FI;k}, \ldots), \quad (h^a_{FF;k}, h^a_{FI;k}, \ldots), \quad k = 1, 2.$$

Here the indexes $k = 1, 2$ denote the first and second part of the game and $FI, \ldots, II$ combinations of choices of Alice and Bob.[1] The upper indexes a, b are marks for Alice's and Bob's payoffs. It is natural to assume that

$$h^b_{FF;1}, h^b_{II;1} > 0, \quad h^b_{FI;1}, h^b_{IF;1} < 0$$

as well as

$$h^a_{FI;1}, h^a_{FI;1} > 0, \quad h^a_{FF;1}, h^a_{II;1} < 0.$$

In the zero sum game

$$h^b_{FF;k} = -h^a_{FF;k}, \quad \ldots, \quad h^b_{II;k} = -h^a_{II;k}.$$

---

[1] We also remark the Alice's choice can be considered as an "element of reality", since her, e.g., F, is really French wine, but Bob's F may be in reality either French or Italian wine, cf. with discussions about realism in quantum mechanics, e.g., [87, 88, 277].

Each part of this game can be represented as an *extensive form game*, hence, by a tree, see http://en.wikipedia.org. This tree is very simple and it has the following branches representing actions of Alice and Bob; each branch ends with the pair of payoffs, the symbol "v" used for vertices and "act" for corresponding actions. The first part of the game is represented by the tree with the branches

$$v = A - - - \text{act} = F - - - v = B - - - \text{act} = F - -(h^b_{FF;1}, h^a_{FF;1});$$

$$v = A - - - \text{act} = F - - - v = B - - - \text{act} = I - -h^b_{FI;1}, h^b_{FI;1});$$

$$v = A - - - \text{act} = I - - - v = B - - - \text{act} = F - -(h^b_{IF;1}, h^a_{IF;1});$$

$$v = A - - - \text{act} = I - - - v = B - - - \text{act} = I - - - (h^b_{II;1}, h^a_{II;1}).$$

As always, we are interested in averages of wins-losses of Alice and Bob.

We consider the following probabilities:

1) Probabilities of Alice's preferences for a bottle of French wine and a bottle of Italian wine, respectively, from the wine collection of the restaurant:

$$p^a_C(F), \quad p^a_C(I).$$

Here the index C is related to the whole context of the game, in particular, to the collection of wines. Another restaurant has another collection of wines, and Alice would have other preferences.

2) Probabilities of recognizing French wine after testing (by Bob) a bottle of French wine that was chosen by Alice for the test: $p^{b|a}_{F|F}$; the probability of a mistake under this condition, i.e., claiming that the wine is Italian, is then $p^{b|a}_{I|F} = 1 - p^{b|a}_{F|F}$. In a similar way we introduce probabilities $p^{b|a}_{I|I}$ and $p^{b|a}_{F|I}$. We have the matrix of transition probabilities

$$\mathbf{P}^{b|a} = (p^{b|a}_{\beta|\alpha}), \quad \beta, \alpha = I, F.$$

Since in this chapter we will consider various combinations of pairwise conditioning, it is convenient to use the upper index to point out precisely the type of conditioning. So, we write $p^{b|a}_{\beta|\alpha}$, instead of $p_{\beta|\alpha}$ (as we did before).

3) Similarly we introduce probabilities $p^b_C(F)$ and $p^b_C(I)$ for Bob's preferences (for the same collection of wines) as well as probabilities $p^{a|b}_{F|F}, \ldots, p^{a|b}_{I|I}$, which represent Alice's ability to recognize the origin of the wine. The matrix of transition probabilities is $\mathbf{P}^{a|b} = (p^{a|b}_{\alpha|\beta})$, $\alpha, \beta = I, F$.

It is convenient to describe this game by a contextual probabilistic model with probabilities $\{p^a_C(\alpha), p^b_C(\beta), p^{b|a}_{\beta|\alpha}), p^{a|b}_{\alpha|\beta}\}$, where $C \in \mathcal{C}$ and the family $\mathcal{C}$ describes the variety of contexts (e.g., wine collections) chosen for the game. We denote this model by the symbol M_{WG}.

Finally, we introduce probabilities that Bob will announce the result $\beta(= F, I)$ in the game that Alice starts with the result α (which is hidden from Bob): $p^{ab}_C(\alpha, \beta) =$

$p_C^a(\alpha)p_{\beta|\alpha}^{b|a}$, and similar probabilities for the game which is started by Bob, $p_C^{ba}(\beta,\alpha)$. The latter are defined via $a|b$-contextual probabilities $p_{\alpha|\beta}^{a|b}$. These are "joint probability distributions", see (3.7).

Probability p_C^{ab} serves well for the first part of the game – when Alice chooses a bottle: $p_C^a(\alpha) = \sum_\beta p_C^{ab}(\alpha,\beta)$. However, it could not be used in the second part of the game, since in general: $p_C^b(\beta) \neq \sum_\alpha p_C^{ab}(\alpha,\beta)$. The second part of the game is served by the probability p_C^{ba}. The tricky thing is really the combination of two games. We point out that in general the equality

$$p_C^{ab}(\alpha,\beta) = p_C^{ba}(\beta,\alpha) \tag{9.1}$$

can be violated, see Sect. 3.1.4. This is the main source of "nonclassicality" of our game.

Then average for wins–losses in the first part of the game for Bob is given by

$$E_1^b(C) = h_{FF;1}^b \; p_C^{ab}(F,F) + h_{FI;1}^b \; p_C^{ab}(F,I) + h_{IF;1}^b \; p_C^{ab}(I,F) + h_{II;1}^b \; p_C^{ab}(I,I),$$

see (3.8). The average for Alice in the first part of the game (in general we can consider a nonzero-sum game) is given by

$$E_1^a(C) = h_{FF;1}^a \; p_C^{ab}(F,F) + h_{FI;1}^a \; p_C^{ab}(F,I) + h_{IF;1}^a \; p_C^{ab}(I,F) + h_{II;1}^a \; p_C^{ab}(I,I).$$

We remark that the maps $(\alpha,\beta) \rightarrow h_{\alpha,\beta;1}^b$ and $(\alpha,\beta) \rightarrow h_{\alpha,\beta;1}^a$ can be considered as classical random variables on the Kolmogorov probability space $\mathcal{P}_C^{ab}$ with $\Omega = X_a \times X_b$, where $X_a = X_b = \{F,I\}$, and probability $\mathbf{P}(\alpha,\beta) = p_C^{ab}(\alpha,\beta)$.

In the same way the averages for Alice and Bob in the second part of the game are given by

$$E_2^a(C) = h_{FF;2}^a \; p_C^{ba}(F,F) + h_{FI;2}^a \; p_C^{ba}(F,I) + h_{IF;2}^a \; p_C^{ba}(I,F) + h_{II;2}^a \; p_C^{ba}(I,I).$$

$$E_2^b(C) = h_{FF;2}^b \; p_C^{ba}(F,F) + h_{FI;2}^b \; p_C^{ba}(F,I) + h_{IF;2}^b \; p_C^{ba}(I,F) + h_{II;2}^b \; p_C^{ba}(I,I).$$

Here the maps $(\beta,\alpha) \rightarrow h_{\beta,\alpha;2}^a$ and $(\beta,\alpha) \rightarrow h_{\beta,\alpha;2}^b$ can be considered as classical random variables on the Kolmogorov probability space $\mathcal{P}_C^{ba}$ with $\Omega = X_b \times X_a$ and probability $\mathbf{P}(\beta,\alpha) = p_C^{ba}(\beta,\alpha)$.

Thus to describe both parts of the game one needs two Kolmogorov probability spaces for fixed context C. To describe games played for a variety of contexts one needs a collection of Kolmogorov spaces. It is more convenient to operate in the contextual probabilistic model M_{WG}.

Averages of total wins–losses are

$$E^b(C) = E_1^b(C) + E_2^b(C), \quad E^a(C) = E_1^a(C) + E_2^a(C).$$

It is convenient to introduce a "wine-observable" for Alice: $a = F, I$. This observable appears in two different contexts. The first context, C, is the context of selection of a bottle from the wine collection. Alice chooses a bottle and says to herself (not Bob!) or just thinks – it is French wine (or it is Italian wine). The second context appears in the second part of the game when Alice should test wine proposed by Bob and after that say: it is French wine (or it is Italian wine). In fact, to be completely correct one should consider two different observables corresponding to these contexts. However, to have a closer analogy with quantum mechanics, we proceed with one observable. Alice is considered as simply an apparatus that says either "French wine" or "Italian wine" (cf. with Stern–Gerlach magnet, which "says" either "spin up" or "spin down"). We remark that our cognitive example shows that it might be more natural to associate with each quantum state – wave function – its own spin-observable. We introduce a similar observable for Bob, $b = F, I$.

9.3 Extensive Form Game with Imperfect Information

As was mentioned, formally the wine testing game is an extensive form game. However, we should point out one rather delicate feature of the game. We recall that a complete extensive form representation specifies: 1) the players of a game; 2) for every player every opportunity they have to move; 3) what each player can do at each of their moves; 4) what each player knows for every move; 5) the payoffs received by every player for every possible combination of moves.

Our game fulfills all those conditions except the fourth one. In fact, the action of Alice does not specify for Bob the result of her action, Bob should guess the country of origin of the wine served by Alice. To come to a conclusion, he should perform a rather complicated analysis of the wine test. One may say that this is a game with *imperfect information.*

We recall that an *information set* is a set of decision nodes such that: 1) every node in the set belongs to one player; 2) when play reaches the information set, the player with the move cannot differentiate between nodes within the information set, i.e. if the information set contains more than one node, the player to whom that set belongs does not know which node in the set has been reached.

If a game has an information set with more than one member, that game is said to have imperfect information. A game with perfect information is such that at any stage of the game, every player knows exactly what has taken place earlier in the game, i.e. every information set is a singleton set. Any game without perfect information has imperfect information.

However, there is a problem with the second condition determining the information set. Of course, Bob does not know precisely which kind of wine is presented for the test. In this sense the set of Bob's nodes after Alice's action (we consider the first part of the game) forms an information set. But (and this is crucial) Bob has the possibility to analyze wine (cf. with measurement process in quantum physics). Therefore he might distinguish two actions of Alice, F and I, but only *partially.* I have no idea whether such a problem of analysis of actions of the opposite player has been discussed in game theory.

9.3.1 Quantum-like Representation of the Wine Testing Game

The wine testing game has a natural QL representation. Let us consider a game with restrictions on probabilities of strategies that make it possible to apply QLRA, see Sect. 4.2: **DS, PO, RC**. By applying QLRA to statistical data we can construct a probability amplitude $\psi_C(\beta)$, $\beta = F, I$. By the condition **RC** such a context C is trigonometric and the probability amplitude is complex valued. It can also be represented by a unit vector of the two-dimensional complex Hilbert space. We remark that in principle there are no reasons for such an assumption. Unlike conventional quantum games, the Wine Game may produce hyperbolic probability amplitudes.

In the trigonometric case we can represent the wins–losses averages in the QL way:

$$E^b(C) = h^b_{FF;1} \, |\langle \psi_C, e^a_F \rangle|^2 \, |\langle e^b_F, e^a_F \rangle|^2 + h^b_{IF;1} \, |\langle \psi_C, e^a_I \rangle|^2 \, |\langle e^b_F, e^a_I \rangle|^2$$

$$+ h^b_{FI;1} \, |\langle \psi_C, e^a_F \rangle|^2 \, |\langle e^b_I, e^a_F \rangle|^2 + h^b_{II;1} \, |\langle \psi_C, e^a_I \rangle|^2 \, |\langle e^b_I, e^a_I \rangle|^2$$

$$+ h^b_{FF;2} \, |\langle \psi_C, e^b_F \rangle|^2 \, |\langle e^b_F, e^a_F \rangle|^2 + h^b_{IF;2} \, |\langle \psi_C, e^b_I \rangle|^2 \, |\langle e^b_I, e^a_F \rangle|^2$$

$$+ h^b_{FI;2} \, |\langle \psi_C, e^b_F \rangle|^2 \, |\langle e^b_F, e^a_I \rangle|^2 + h^b_{II;2} \, |\langle \psi_C, e^b_I \rangle|^2 \, |\langle e^b_I, e^a_I \rangle|^2.$$

In the same way we represent the average for Alice. Thus the wine testing game satisfying conditions **DS, PO, RC** can be represented in the complex Hilbert space.

The QL expression for the average is significantly simpler in the case of a zero sum game with symmetry between the first and second parts: $h^b_{FF;1} = h^a_{FF;2} = -h^b_{FF;2}, \ldots, h^b_{II;1} = h^a_{II;2} = -h^b_{II;2}$. Here

$$E^b(C) = h^b_{FF;1}(|\langle \psi_C, e^a_F \rangle|^2 \, |\langle e^b_F, e^a_F \rangle|^2 - |\langle \psi_C, e^b_F \rangle|^2 \, |\langle e^b_F, e^a_F \rangle|^2)$$

$$+ h^b_{IF;1}(|\langle \psi_C, e^a_I \rangle|^2 \, |\langle e^b_F, e^a_I \rangle|^2 - |\langle \psi_C, e^b_I \rangle|^2 \, |\langle e^b_I, e^a_F \rangle|^2)$$

$$+ h^b_{FI;1}(|\langle \psi_C, e^a_F \rangle|^2 \, |\langle e^b_I, e^a_F \rangle|^2 - |\langle \psi_C, e^b_F \rangle|^2 \, |\langle e^b_F, e^a_I \rangle|^2)$$

$$+ h^b_{II;1}(|\langle \psi_C, e^a_I \rangle|^2 \, |\langle e^b_I, e^a_I \rangle|^2 - |\langle \psi_C, e^b_I \rangle|^2 \, |\langle e^b_I, e^a_I \rangle|^2)$$

$$= h^b_{FF;1} \, |\langle e^b_F, e^a_F \rangle|^2 \, (|\langle \psi_C, e^a_F \rangle|^2 - |\langle \psi_C, e^b_F \rangle|^2) + h^b_{IF;1} \, |\langle e^b_F, e^a_I \rangle|^2$$

$$(|\langle \psi_C, e^a_I \rangle|^2 - |\langle \psi_C, e^b_I \rangle|^2)$$

$$+ h^b_{FI;1} \, |\langle e^b_I, e^a_F \rangle|^2 \, (|\langle \psi_C, e^a_F \rangle|^2 - |\langle \psi_C, e^b_F \rangle|^2) + h^b_{II;1} \, |\langle e^b_I, e^a_I \rangle|^2$$

$$(|\langle \psi_C, e^a_I \rangle|^2 - |\langle \psi_C, e^b_I \rangle|^2)$$

$$= (|\langle \psi_C, e^a_F \rangle|^2 - |\langle \psi_C, e^b_F \rangle|^2) \, (h^b_{FF;1} \, |\langle e^b_F, e^a_F \rangle|^2 + h^b_{FI;1} \, |\langle e^b_I, e^a_F \rangle|^2)$$

$$+ (|\langle \psi_C, e^a_I \rangle|^2 - |\langle \psi_C, e^b_I \rangle|^2) \, (h^b_{IF;1} \, |\langle e^b_F, e^a_I \rangle|^2 + h^b_{II;1} \, |\langle e^b_I, e^a_I \rangle|^2).$$

9.3.2 Superposition of Preferences

We now point out that we can expand e.g. vectors of the b-basis with respect to the a-basis: $e_F^b = c_{FF}e_F^a + c_{FI}e_I^a$, $e_I^b = c_{IF}e_F^a + c_{II}e_I^a$. One might say that "Bob's preferences are superpositions of Alice preferences." However, we cannot assign any real meaning to such a sentence in the present game framework. Thus superposition is merely a purely mathematical representation – the geometric picture of the probabilistic structure of the game. In the same way we can expand the state ψ_C with respect to the a-basis as well as the b-basis. Such expansions do not have any real meaning either, just geometrical representation of probabilities. Nevertheless, such a picture is convenient for geometric representation of mental states of Alice and Bob. One may use the following geometric model: there are two basic mental states of Alice (in the context of the Wine Game) e_F^a and e_I^a. In general Alice plays in the superposition of these states $\psi = c_F^a e_F^a + c_I^a e_I^a$. In the same way Bob has two basic mental states e_F^b and e_I^b. In general Bob plays in the superposition of these states $\psi = c_F^b e_F^b + c_I^b e_I^b$. Moreover, (at least mathematically) Bob's mental states can be represented as superpositions of Alice's mental states.

We can represent the average of Bob's wins–losses in the interference form

$$
\begin{aligned}
E^b(C) = & (|\langle \psi_C, e_F^a \rangle|^2 - |\bar{c}_{FF} \langle \psi_C, e_F^a \rangle + \bar{c}_{FI} \langle \psi_C, e_I^a \rangle|^2) (h_{FF;1}^b |\langle e_F^b, e_F^a \rangle|^2 \\
& +h_{FI;1}^b |\langle e_I^b, e_F^a \rangle|^2) \\
& +(|\langle \psi_C, e_I^a \rangle|^2 - |\bar{c}_{IF} \langle \psi_C, e_F^a \rangle + \bar{c}_{II} \langle \psi_C, e_I^a \rangle|^2) (h_{IF;1}^b |\langle e_F^b, e_I^a \rangle|^2 \\
& +h_{II;1}^b |\langle e_I^b, e_I^a \rangle|^2) \\
= & (|\langle \psi_C, e_F^a \rangle|^2 - (|\langle \psi_C, e_F^a \rangle|^2 |\langle e_F^b, e_F^a \rangle|^2 + |\langle \psi_C, e_I^a \rangle|^2 |\langle e_F^b, e_I^a \rangle|^2 \\
& +2\cos\theta |\langle \psi_C, e_F^a \rangle \langle e_F^b, e_F^a \rangle \langle \psi_C, e_I^a \rangle \langle e_F^b, e_I^a \rangle|)) (h_{FF;1}^b |\langle e_F^b, e_F^a \rangle|^2 \\
& +h_{FI;1}^b |\langle e_I^b, e_F^a \rangle|^2) \\
& +(|\langle \psi_C, e_I^a \rangle|^2 - (|\langle \psi_C, e_F^a \rangle|^2 |\langle e_I^b, e_F^a \rangle|^2 + |\langle \psi_C, e_I^a \rangle|^2 |\langle e_I^b, e_I^a \rangle|^2 \\
& -2\cos\theta |\langle \psi_C, e_F^a \rangle \langle e_I^b, e_F^a \rangle \langle \psi_C, e_I^a \rangle \langle e_I^b, e_I^a \rangle|)) (h_{IF;1}^b |\langle e_F^b, e_I^a \rangle|^2 \\
& +h_{II;1}^b |\langle e_I^b, e_I^a \rangle|^2).
\end{aligned}
$$

9.3.3 Interpretation of Gambling Wave Function

The wave function ψ_C was constructed on the basis of probabilities: $p_C^a(\alpha)$, $p_C^a(\beta)$, $p_{\beta|\alpha}^{b|a}$. It represents the wine collection of the restaurant as well as preferences of Alice and Bob. Moreover, it also represents their abilities to find the difference between French and Italian wines.

Thus one may say such a wave function (complex probability amplitude) ψ_C has no real counterpart. We could not point to any object in reality that is represented by ψ_C. It represents the context of the wine collection as well as Bob's and Alice's preferences and experiences with different kinds of wines. Such a context is extremely complex. It is impossible to describe its precisely. However, the ψ_C provides some approximative representation of this context in the complex Hilbert space.

We note that Alice and Bob are coupled through the wave function. The wave function really provides a way to combine probabilistic features of two cognitive systems, Alice and Bob, which could not be incorporated into a single Kolmogorov probability space.

9.3.4 The Role of Bayes Formula

Suppose for the moment that the randomness of actions of Alice and Bob can be described by the Kolmogorov probability space $\mathcal{P} = (\Omega, \mathcal{F}, \mathbf{P})$ in the following way:

a) The wine-collection context C is represented by an element of $\mathcal{F}$, which will be denoted by the same symbol.
b) Probabilities

$$p_C^a(\alpha) = \mathbf{P}_C(C_\alpha) = \frac{\mathbf{P}(C_\alpha \cap C)}{\mathbf{P}(C)}, \quad p_C^b(\beta) = \mathbf{P}_C(C_\beta) \equiv \frac{\mathbf{P}(C_\beta \cap C)}{\mathbf{P}(C)},$$

where

$$C_\alpha = \{\omega \in \Omega : a(\omega) = \alpha\}, \quad C_\beta = \{\omega \in \Omega : b(\omega) = \beta\}.$$

Here $\mathbf{P}_C$ is the conditional probability measure corresponding to the subset C of $\mathcal{F} : \mathbf{P}_C(A) = \mathbf{P}(A \cap C)/\mathbf{P}(C)$.
c) The $b|a$-contextual probabilities

$$p_{\beta|\alpha}^{b|a} = \mathbf{P}_C(C_\beta|C_\alpha) \equiv \frac{\mathbf{P}(C_\beta \cap C_\alpha \cap C)}{\mathbf{P}(C_\alpha \cap C)}.$$

In such a representation the C-conditional Bayes formula holds:

$$\mathbf{P}_C(C_\alpha \cap C_\beta) = \mathbf{P}_C(C_\alpha)\mathbf{P}_C(C_\beta|C_\alpha). \tag{9.2}$$

Hence, here the equality (9.1) holds! (Because the Kolmogorovian probability is symmetric: $\mathbf{P}_C(C_\alpha \cap C_\beta) = \mathbf{P}_C(C_\beta \cap C_\alpha)$.) We obtain the following equality:

$$\mathbf{P}_C(C_\alpha)\mathbf{P}_C(C_\beta|C_\alpha) = \mathbf{P}_C(C_\beta)\mathbf{P}_C(C_\alpha|C_\beta). \tag{9.3}$$

Since we want the QL representation, we consider symmetrically conditioned variables a and b, see Sect. 3.1.3 – so the condition **SC** holds: (2.18). The condition (9.3) implies that

$$\mathbf{P}_C(C_\alpha) = \mathbf{P}_C(C_\beta) = 1/2. \tag{9.4}$$

Thus one can construct a Kolmogorov representation of the Wine Game satisfying conditions a)–c) iff selection of wines from the collection is uniformly distributed between French and Italian wines (for both Alice and Bob). If not, then a single Kolmogorov space does not exist. For example, if the probability that Alice chooses a bottle of French wine $p_C^a(F) = 1/3$ (and consequently the probability that she chooses a bottle of Italian wine $p_C^a(I) = 2/3$), then it is impossible to construct a Kolmogorov probability space for this game. Of course, one should not forget that we assumed that the game probabilities are coupled to the Kolmogorov space via conditions a)–c) and that we would like to have symmetric transition probabilities. The origin of this nonclassicality of the probabilistic description is the *impossibility to combine on a single Kolmogorov space preferences of Alice and Bob in choosing wines and their abilities to test wines.* By using Gudder's theory of probability manifolds [126, 127] we can say that we have a probability manifold with the atlas having two charts, one serves for the first part of game and another for the second.

We remark that the choice c) of the $b|a$-contextual probabilities implies immediately that the *coefficients of interference* λ are equal to zero.

9.3.5 Action at a Distance?

One can consider a more advanced Wine Game involving facelogy: Bob can extract some information about the origin of the wine by observing the behavior of Alice after she has made her choice. Using the terminology of quantum mechanics one can talk about "action at a distance." However, even if such action is present in the game it is not instantaneous! Everything happens in complete accordance with the laws of special relativity: light is reflected from Alice's face and Bob obtains information only when the light wave reaches his eyes.

Consideration of QL games of the facelogy-type extends essentially the range of possible applications of our model. However, we do not couple directly such action at a distance with essentially nonclassical probabilistic structure. The origin of nonclassicality is the impossibility of combining all possible preferences in a single probability space.

9.4 Wine Game with Three Players

We now generalize the Wine Game by considering three players, Alice, Bob, Cecilia. The first part: Alice chooses a bottle, Bob tests; the second part: Bob chooses, Cecilia tests, and the third part: Cecilia chooses, Alice tests. We shall use probabilities with indexes a, b, c corresponding to Alice, Bob, Cecilia. For each part of the game we fix payment matrices. We consider a symmetric game. We can write averages: for the first part $E_1^b(C) (= -E_1^a(C))$, for the second part $E_2^c(C) (= -E_2^b(C))$, and for the third part $E_3^a(C) (= -E_3^c(C))$.

We assume that conditions **DS, PO, RC**, Sect. 4.2, which guarantee the possibility of applying QLRA, hold for all pairs of observables. Thus we apply QLRA to the probabilities corresponding to the pair a, b. We obtain the complex probability amplitude $\psi_{C;b|a}$, which belongs to the two-dimensional Hilbert space that is denoted $\mathcal{H}_{b|a}$. Observables a, b are represented by self-adjoint operators $\hat{a}, \hat{b}$, which have bases of eigenvectors $\{e_\alpha^{a;b|a}\}, \{e_\beta^{b;b|a}\}$. We also apply QLRA to the probabilities corresponding to the pair b, c. We obtain a new complex probability amplitude $\psi_{C;c|b}$, which belongs to the two-dimensional Hilbert space that is denoted $\mathcal{H}_{c|b}$. Observables b, c are represented by self-adjoint operators $\hat{b}, \hat{c}$, which have bases of eigenvectors $\{e_\beta^{b;c|b}\}, \{e_\gamma^{c;c|b}\}$. Finally, consider the $\mathcal{H}_{a|b}$-representation.

These representations can be identified with the aid of unitary maps:

$$U_{b|a}^{c|b} : \mathcal{H}_{b|a} \rightarrow \mathcal{H}_{c|b}, \quad e_\beta^{b;b|a} \rightarrow e_\beta^{b;c|b},$$

and

$$U_{c|b}^{a|c} : \mathcal{H}_{c|b} \rightarrow \mathcal{H}_{a|c}, \quad e_\gamma^{c;c|b} \rightarrow e_\gamma^{c;a|c}.$$

The crucial point is that $U_{b|a}^{c|b}(\psi_{C;b|a}) = \psi_{C;c|b}$ and $U_{c|b}^{a|c}(\psi_{C;c|b}) = \psi_{C;a|c}$. Therefore we can identify complex probability amplitudes $\psi_{C;b|a}, \psi_{C;c|b}, \psi_{C;a|c}$ and consider a unit vector ψ_C as representing the wine collection and preferences of Alice, Bob and Cecilia. We shall come back to this game little bit later.

We remark that this game has the structure of Gudder's probability manifold with the atlas having three charts.

9.5 Simulation of the Wine Game

Typically quantum probabilities are imagined as rather mysterious things. Absence of the underlying Kolmogorov space may only support such a viewpoint. However, by using the frequency (von Mises) approach quantum probabilities can be easily simulated. One need not use special "quantum coins" given by sources of photons or electrons. We simulate our game by using the following system of dichotomous random generators (taking values F and I):

$$g_a, \quad g_b, \quad g^{b|a}(\alpha), \quad g^{a|b}(\beta).$$

Here g_a and g_b simulate choices of wine from the collection C (by Alice and Bob, respectively); the frequencies of F and I approach the corresponding probabilities $p_C^a(F), p_C^a(I), p_C^b(F), p_C^b(I)$ when the number of trials goes to infinity. The generator $g^{b|a}(\alpha)$ describes the ability of Bob to analyze the wine's origin under the condition that Alice selects a bottle of α-origin. For example, the generator $g^{b|a}(F)$ takes the value F if Bob correctly recognized French wine (which was chosen by Alice). The generator $g^{a|b}(\beta)$ has a similar meaning.

Now, to simulate the Wine Game, we just apply these generators consequently in the right order, e.g., first g_a and if it takes the value F, then the generator $g^{b|a}(F)$. That's all! We shall simulate probabilities and payoffs given by the two-dimensional QL model.

In all previous considerations we started with some collection of probabilities and transition probabilities, and under the conditions **DS, PO, RC** we were able to represent the Wine Game in the two-dimensional complex Hilbert space. By applying QLRA we constructed the wave function and operators $\hat{a}$, $\hat{b}$.

We can also proceed in the opposite way. We can take two noncommutative operators in the two-dimensional Hilbert space, say $\hat{a}$ and $\hat{b}$, and a normalized vector ψ in this space. Then we find (by using Born's rule) all the probabilities that we need for the Wine Game. Those probabilities will automatically satisfy conditions **DS, PO, RC**. Finally, we can simulate the Wine Game by using the above scheme.

This strategy is especially convenient for generalizations of the Wine Game to spaces of high dimension. QLRA becomes very complicated, see [214]. Reconstruction of the wave function is not a simple task. Therefore one can start just with probabilities, which are obtained from the mathematical formalism of quantum mechanics. Moreover, the possibility of applying QLRA is restricted by a number of conditions. One can ignore these conditions by starting directly with a normalized vector ψ.

9.6 Bell's Inequality for Averages of Payoffs

We now come back to the Wine Game with three players, Alice, Bob, Cecilia. We shall use the pragmatic strategy proposed at the end of the previous section. We take probabilities and operators corresponding to a known quantum system and simulate the Wine Game on the basis of these probabilities. We emphasize that we take probabilities given by the mathematical apparatus of quantum mechanics and not at all a quantum physical system by itself. We introduce a game parameter $\theta \in [0, 2\pi)$. Alice is characterized by $\theta = \theta_1$, Bob by $\theta = \theta_2$, Cecilia by $\theta = \theta_3$. We take the transition probabilities corresponding to the "spin-1/2 system." For the first part of the game we have

$$p_{F|F}^{b|a} = p_{I|I}^{b|a} = \cos^2 \frac{\theta_1 - \theta_2}{2};$$

$$p_{I|F}^{b|a} = p_{F|I}^{b|a} = \sin^2 \frac{\theta_1 - \theta_2}{2}.$$

The "transition probabilities" for other parts of the game are defined in a similar way, e.g.,

$$p^{c|b}_{F|F} = p^{c|b}_{I|I} = \cos^2 \frac{\theta_2 - \theta_3}{2}.$$

Let us choose the following matrix of payoffs:

$$h_{FF} = h_{II} = +1, \quad h_{IF} = h_{FI} = -1.$$

Let us now suppose that Alice, Bob and Cecilia select wine from the collection by using uniform random generators: $p^a_C(\alpha) = p^b_C(\beta) = p^c_C(\gamma) = 1/2$. We now find the average for Bob's wins–losses in the first part of the Wine Game:

$$E^b_1 \equiv E(\theta_1, \theta_2) = \cos^2 \frac{\theta_1 - \theta_2}{2} - \sin^2 \frac{\theta_1 - \theta_2}{2} = \cos(\theta_1 - \theta_2). \quad (9.5)$$

In the same way we have for Cecilia

$$E^c_2 \equiv E(\theta_2, \theta_3) = \cos(\theta_2 - \theta_3) \quad (9.6)$$

and finally for Alice

$$E^a_3 \equiv E(\theta_3, \theta_1) = \cos(\theta_3 - \theta_1). \quad (9.7)$$

We set now $F = +1$ and $I = -1$. In this notation both observables a and b take the values $+1$. We proceed inside the contextual probabilistic model M_{WG}. Averages, see (3.4) in Sect. 3.1.1, $\bar{a}_C$ and $\bar{b}_C$ are equal to zero. Thus averages for wins–losses are nothing other than corresponding covariances, see (3.9) in Sect. 3.1.4:

$$E^b_1 = \text{cov}_C(a, b), \quad E^c_2 = \text{cov}_C(c, b), \quad E^a_3 = \text{cov}_C(c, a).$$

These covariances are taken with respect to probabilities p^{ab}_C, p^{bc}_C, p^{ca}_C. We now ask:

Can one construct a probability measure **P** *and realize observables* a, b, c *by random variables on the corresponding Kolmogorov space in such a way that*

$$\mathbf{P}(a = \alpha, b = \beta) = p^{ab}_C(a = \alpha, b = \beta), \quad \mathbf{P}(b = \beta, c = \gamma) = p^{bc}_C(b = \beta, c = \gamma),$$
$$(9.8)$$
$$\mathbf{P}(c = \gamma, a = \alpha) = p^{ca}_C(c = \gamma, a = \alpha)? \quad (9.9)$$

The answer is negative. If representations (9.8), (9.9) can be constructed, then one can prove Bell's inequality, see Sect. 2.2.2, Theorem 2.2:

$$|\text{cov}_C(a, b) - \text{cov}_C(b, c)| \le 1 - \text{cov}_C(c, a). \quad (9.10)$$

Bell's inequality is violated for covariances given by (9.5)–(9.7) for some choices of parameters.

If there were a classical probabilistic model behind the Wine Game (for some set of probabilities), then averages of payments would satisfy the following Bell's inequality:

$$|E_1^b - E_2^c| \le 1 - E_3^a. \qquad (9.11)$$

Even intuitively it is clear that there are no reasons to assume that this inequality should hold for any set of probabilities.

The expression on the left-hand side of the Bell's inequality is equal to the average of the total wins–losses of Bob in the game (i.e., in the two series of games – with Alice and Cecilia, in the first Bob tests wine and the second Cecilia does this): $E^b = E_1^b + E_2^b = E_1^b - E_2^c = \cos(\theta_1 - \theta_2) - \cos(\theta_2 - \theta_3)$.

Finally, we remark that the Wine Game can be generalized to the Hilbert space H of an arbitrary dimension. The only difference is that now the collection C contains wines from n countries, which are labeled by $i = 1, \ldots, n$.

Let us consider two self-adjoint operators $\hat{a}$ and $\hat{b}$ and the corresponding orthonormal bases of eigenvectors $\{e_i^a\}_{i=1}^n$ and $\{e_j^b\}_{j=1}^n$. In principle, operators could even commute. Of course, QLRA would not work in such a case. But our task is not to reconstruct probabilities from the game, but only to simulate the game.

We also take a normalized vector $\psi \in \mathcal{H}$. This vector ψ describes collections of wines created by Alice and Bob as well as their experiences of testing of wines. Actions of Alice and Bob are now labeled by $i = 1, \ldots, n$. The tree of this extensive form game has n nodes leaving this vertex. Bob's average is given by

$$E^b = \sum_{i,j=1}^n h_{ji;1}^b \, |\langle \psi_C, e_j^a \rangle \langle e_j^b, e_j^a \rangle|^2 + \sum_{i,j=1}^n h_{ij;2}^b \, |\langle \psi_C, e_i^b \rangle \langle e_i^b, e_j^a \rangle|^2.$$

Chapter 10
Psycho-financial Model

This chapter presents a model that was proposed by Olga Choustova and me [53–62, 175]. She applied methods of quantum mechanics to mathematical modeling of price dynamics in the financial market. She pointed out that behavioral financial factors (e.g., expectations of traders) could be described by using the pilot wave (Bohmian) model of quantum mechanics; see Section 12.6 for a brief introduction to the mathematical formalism of Bohmian mechanics.

Trajectories of prices are determined by two financial potentials: classical-like $V(t, q)$ ("hard" market conditions, e.g., natural resources) and QL $U(t, q)$ (behavioral market conditions).

On the one hand, our Bohmian model is a QL model for the financial market, cf. publications of Segal and Segal [274], Baaquie [22, 23], Haven [132–135], Piotrowski et al. [252–258], and Danilov and Lambert-Mogiliansky [73–76].

On the other hand (since Bohmian mechanics provides the possibility of describing individual price trajectories), it belongs to the domain of extensive research on deterministic dynamics for financial assets (Granger [122], Barnett and Serletis [28], Benhabib [34], Brock and Sayers [43], Campbell et al. [52], Hsieh [152], and many others).

10.1 Deterministic and Stochastic Models of Financial Markets

10.1.1 Efficient Market Hypothesis

In economics and financial theory, analysts use random walks and more general martingale techniques to model the behavior of asset prices, in particular share prices on stock markets, currency exchange rates and commodity prices. This practice has its basis in the presumption that investors act *rationally and without bias,* and that at any moment they estimate the value of an asset based on future expectations. Under these conditions, all existing information affects the price, which changes only when new information comes out. By definition, *new information appears randomly and influences the asset price randomly.* Corresponding continuous time models are based on stochastic processes (this approach was initiated in the thesis

of Bachelier [24] in 1890), see, e.g., the books of Mantegna and Stanley [237] and
Shiryaev [281] for historical and mathematical details.

This practice was formalized through the *efficient market hypothesis*, which was
formulated in the 1960s, see Samuelson [269, 270] and Fama [103] for details:

*A market is said to be efficient in the determination of the most rational price if
all the available information is instantly processed when it reaches the market and
it is immediately reflected in a new value of prices of the assets traded.*

The efficient market hypothesis was supported by statistical investigations of
Samuelson [269]. Mathematically the efficient markets hypothesis means that financial markets can be described by classical stochastic processes and they are of a very
special type, namely, so-called martingales.

10.1.2 Deterministic Models for Dynamics of Prices

First we remark that empirical studies have demonstrated that prices do not completely follow a random walk. Low serial correlations (around 0.05) exist in the
short term and slightly stronger correlations over the longer term. Their sign and the
strength depend on a variety of factors, but transaction costs and bid–ask spreads
generally make it impossible to earn excess returns. Interestingly, researchers have
found that some of the biggest price deviations from a random walk result from
seasonal and temporal patterns, see [237].

There are also a variety of arguments, both theoretical and obtained on the basis
of statistical analysis of data, that question the general martingale model (and hence
the efficient market hypothesis), see, e.g., [122, 28, 34, 52, 152]. It is important
to note that efficient markets imply there are no exploitable profit opportunities. If
this is true then trading on the stock market is a game of chance and not of any
skill, but traders buy assets they think are undervalued in the hope of selling them at
their true price for a profit. If market prices already reflect all available information,
then where does the trader draw this privileged information from? Since there are
thousands of very well informed, well-educated asset traders, backed by many data
researchers, buying and selling securities quickly, logically assets markets should
be very efficient and profit opportunities should be minimal. On the other hand,
we see that there are many traders who successfully use their opportunities and
continuously carry out very successful financial operations; see the book by Soros
[282] for discussion.[1] Intensive investigations testing whether real financial data
can be really described by the martingale model, have also been performed, see
[122, 28, 34, 52, 152]. Roughly speaking, people try to understand the following on
the basis of available financial data:

Do financial asset returns behave randomly (hence unpredictably) or deterministically (in which case one may hope to predict them and even to construct a
deterministic dynamical system which would at least mimic the dynamics of the
financial market)?

---

[1] It seems that Soros is sure he does not work at efficient markets.

Predictability of financial asset returns is a broad and very active research topic and a complete survey of the vast literature is beyond the scope of this work. We note, however, that there is a rather general opinion that financial asset returns are predictable, see [122, 28, 34, 52, 152].

10.1.3 Behavioral Finance and Economics

We point out that there is no general consensus on the validity of the efficient market hypothesis. As is pointed out in [52]: "...econometric advances and empirical evidence seem to suggest that financial asset returns are predictable to some degree. Thirty years ago this would have been tantamount to an outright rejection of market efficiency. However, modern financial economics teaches us that other, perfectly rational factors may account for such predictability. The fine structure of securities markets and frictions in trading process can generate predictability. Time-varying expected returns due to changing business conditions can generate predictability. A certain degree of predictability may be necessary to reward investors for bearing certain dynamic risks."

Therefore it would be natural to develop approaches which are not based on the assumption that investors act *rationally and without bias* and that, consequently, new information appears randomly and influences the asset price randomly. In particular, there are two well established (and closely related) fields of research, *behavioral finance and behavioral economics*, which apply scientific research on human and social cognitive and emotional biases[2] to better understand economic decisions and how they affect market prices, returns and the allocation of resources. The fields are primarily concerned with the rationality, or lack thereof, of economic agents. Behavioral models typically integrate insights from psychology with neo-classical economic theory. Behavioral analysis is mostly concerned with the effects of market decisions, but also those of public choice, another source of economic decisions with some similar biases.

Since the 1970s, the intensive exchange of information in the world of finances has become one of the main factors determining the dynamics of prices. Electronic trading (which has become the most important part of the environment of the major stock exchanges) induces huge information flows between traders (including the foreign exchange market). Financial contracts are made on a new time scale that differs essentially from the old "hard" time scale that was determined by the development of the economic basis of the financial market. Prices at which traders are willing

---

[2] Cognitive bias is any of a wide range of observer effects identified in cognitive science, including very basic statistical and memory errors that are common to all human beings and drastically skew the reliability of anecdotal and legal evidence. They also significantly affect the scientific method, which is deliberately designed to minimize such bias from any one observer. They were first identified by Amos Tversky and Daniel Kahneman as a foundation of behavioral economics, see, e.g., [293]. Bias arises from various life, loyalty and local risk and attention concerns that are difficult to separate or codify. Tversky and Kahneman claim that they are at least partially the result of problem-solving using heuristics, including the availability heuristic and the representativeness.

to buy (bid quotes) or sell (ask quotes) a financial asset are not only determined by the continuous development of industry, trade, services, the situation on the market of natural resources and so on. Information (mental, market-psychological) factors play a very important (and in some situations crucial) role in price dynamics. Traders performing financial operations work as a huge collective cognitive system. Roughly speaking, classical-like dynamics of prices (determined) by "hard" economic factors are permanently perturbed by additional financial forces, mental (or market-psychological) forces, see the book [282].

10.1.4 Quantum-like Model for Behavioral Finance

Olga Choustova has developed a new approach that is not based on the assumption that investors act rationally and without bias and that, consequently, new information appears randomly and influences the asset price randomly. Her approach can be considered as a special econophysical [237] model in the domain of behavioral finance. In her approach information about the financial market (including expectations of agents of the financial market) is described by an *information field* $\psi(t, q)$ – *a financial wave*. This field evolves deterministically, perturbing the dynamics of prices of stocks and options. The dynamics is given by Schrödinger's equation on the space of prices of shares. Since the psychology of agents of the financial market makes an important contribution to the financial wave $\psi(t, q)$, our model can be considered as a special *psycho-financial model*.

Choustova's model can be also considered as a contribution to applications of quantum mechanics outside the microworld, see [8, 161, 173, 176]. Her model is fundamentally based on investigations by Bohm, Hiley, and Pylkkänen [40, 144] of the *active information* interpretation of Bohmian mechanics and its applications to cognitive sciences, see also [176].

In her model Choustova used methods of Bohmian mechanics to simulate dynamics of prices on the financial market. She started with the development of the classical Hamiltonian formalism on the price/price-change phase space to describe the classical-like evolution of prices. This classical dynamics of prices is determined by "hard" financial conditions (natural resources, industrial production, services and so on). These conditions, as well as "hard" relations between traders at the financial market, are mathematically described by the classical financial potential. At the real financial market "hard" conditions are not the only source of price changes. Information and market psychology play an important (and sometimes determining) role in price dynamics.

Choustova proposed that these "soft" financial factors be described by using the pilot wave (Bohmian) model of quantum mechanics. The theory of financial mental (or psychological) waves was used to take into account market psychology. In Choustova's model the real trajectories of prices are determined (by the financial analogue of the second Newton law) by two financial potentials: classical-like ("hard" market conditions) and QL ("soft" market conditions).

This QL model of financial processes was strongly motivated by consideration by Soros [282] of the financial market as a complex cognitive system. Such an approach he called the theory of *reflexivity*. In this theory there is a large difference between market that is "ruled" by only "hard" economical factors and a market where mental factors play a crucial role (even changing the evolution of the "hard" basis, see [282]).

Soros rightly remarked that the "nonmental" market evolves due to classical random fluctuations. However, such fluctuations do not provide an adequate description of the mental market. He proposed that an analogy with quantum theory be used. However, it was noticed that quantum formalism could not be applied directly to the financial market [282]. Traders differ essentially from elementary particles. Elementary particles behave stochastically due to perturbation effects provided by measurement devices.

According to Soros, traders at the financial market behave stochastically due to free will of individuals. Combinations of a huge number of free wills of traders produce additional stochasticity at the financial market that could not be reduced to classical random fluctuations (determined by nonmental factors). Here Soros followed the conventional (Heisenberg, Bohr, Dirac) view of the origin of quantum stochasticity. However, in the Bohmian approach (that is the nonconventional one) quantum statistics is induced by the action of an additional potential, the quantum potential, that changes classical trajectories of elementary particles. Such an approach provides the possibility of applying quantum formalism to the financial market.

We remark that applications of the pilot-wave theory to financial option pricing were considered by Haven in [136]. There were also numerous investigations on applying quantum methods to the financial market, see, e.g., [132, 135], that were not directly coupled to behavioral modeling, but based on the general concept that randomness of the financial market can be better described by quantum mechanics, see, e.g., Segal and Segal [274]: "A natural explanation for extreme irregularities in the evolution of prices in financial markets is provided by quantum effects." Non-Bohmian quantum models for the financial market (in particular, based on quantum games) were developed by Piotrowski, Sladkowski, and coworkers, see [252, 254]. Some of those models can also be considered as behavioral QL models.

An interesting contribution to behavioral QL modeling is the theory of nonclassical measurements in behavioral sciences (with applications to economics) that was developed by Danilov and Lambert-Mogiliansky [73, 74].

10.2 Classical Econophysical Model of the Financial Market

10.2.1 Financial Phase Space

Let us consider a mathematical model in which a huge number of agents of the financial market interact with one another and take into account external economic (as well as political, social and even meteorological) conditions in order to determine the price to buy or sell financial assets. We consider the trade with shares of some corporations (e.g., Volvo, Saab, Ikea,...).[3]

We consider a *price system of coordinates.* We enumerate corporations that emitted shares in the financial market under consideration: $j = 1, 2, \ldots, n$ (e.g., Volvo: $j = 1$, Saab: $j = 2$, Ikea: $j = 3,\ldots$). Introduce the n-dimensional configuration space $Q = \mathbf{R}^n$ of prices, $q = (q_1, \ldots, q_n)$, where q_j is the price of a share of the jth corporation. Here $\mathbf{R}$ is the real line. The dynamics of prices is described by the trajectory $q(t) = (q_1(t), \ldots, q_n(t))$ in the configuration price space Q.

Another variable under consideration is the *price change variable*:

$$v_j(t) = \dot{q}_j(t) = \lim_{\Delta t \to 0} \frac{q_j(t + \Delta t) - q_j(t)}{\Delta t},$$

see, for example, the book [237] on the role of the price change description. In real models we consider the discrete time scale $\Delta t, 2\Delta t, \ldots$. Here we should use a discrete price change variable $\delta q_j(t) = q_j(t + \Delta t) - q_j(t)$.

We denote the space of price changes (price velocities) by the symbol $V(\equiv \mathbf{R}^n)$ with coordinates $v = (v_1, \ldots, v_n)$. As in classical physics, it is useful to introduce the phase space $Q \times V = \mathbf{R}^{2n}$, namely the *price phase space*. A pair $(q, v) = $ (price, price change) is called the *state of the financial market.*

Later we shall consider QL states of the financial market. The state (q, v) that we consider at the moment is a classical state.

We now introduce an analogue m of mass as the number of items (in our case shares) that a trader emitted to the market.[4] We call m the *financial mass.* Thus each trader j (e.g., Volvo) has its own financial mass m_j (the size of the emission of its shares). The total price of the emission of the jth trader is equal to $T_j = m_j q_j$ (this is nothing other than *market capitalization*). Of course, it depends on time: $T_j(t) = m_j q_j(t)$. To simplify considerations, we consider a market in which any emission of shares is of fixed size, so m_j does not depend on time. In principle, our model can be generalized to describe a market with time-dependent financial masses, $m_j = m_j(t)$.

We also introduce *financial energy* of the market as a function $H : Q \times V \to \mathbf{R}$. Let us use the analogy with classical mechanics. (Why not? In principle, there is not so much difference between motions in "physical space" and "price space".) In this case we could consider (at least for mathematical modeling) the financial energy of the form

---

[3] Similar models can be developed for trade with options, see Haven [136] for the Bohmian financial wave model for portfolios.

[4] 'Number' is a natural number $m = 0, 1, \ldots$, the price of share, e.g., in US dollars. However, in a mathematical model it can be convenient to consider real m. This can be useful for conversions from one currency to another.

$$H(q, v) = \frac{1}{2} \sum_{j=1}^{n} m_j v_j^2 + V(q_1, \ldots, q_n). \tag{10.1}$$

Here $K(q, v) = 1/2 \sum_{j=1}^{n} m_j v_j^2$ is the *kinetic financial energy* and $V(q_1, \ldots, q_n)$ is the potential financial energy; m_j is the financial mass of the jth trader.

The kinetic financial energy represents efforts of agents of the financial market to change prices: higher price changes induce higher kinetic financial energies. If the corporation j_1 has higher financial mass than the corporation j_2, so $m_{j_1} > m_{j_2}$, then the same change of price, i.e., the same financial velocity $v_{j_1} = v_{j_2}$, is characterized by higher kinetic financial energy: $K_{j_1} > K_{j_2}$. We also remark that high kinetic financial energy characterizes rapid changes of the financial situation at the market. However, the kinetic financial energy does not give the sign of these changes. It could be rapid economic growth as well as recession.

The *potential financial energy* V describes the interactions between traders $j = 1, \ldots, n$ (e.g., competition between Nokia and Ericsson) as well as external economic conditions (e.g., the price of oil and gas) and even meteorological conditions (e.g., the weather conditions in Louisiana and Florida). For example, we can consider the simplest interaction potential:

$$V(q_1, \ldots, q_n) = \sum_{j=1}^{n} (q_i - q_j)^2.$$

The difference $|q_1 - q_j|$ between prices is the most important condition for *arbitrage*.

We could never take into account all economic and other conditions that influence the market. Therefore by using some concrete potential $V(t, q)$ we consider a very idealized model of financial processes. However, such an approach is standard for physical modeling, where we also consider idealized mathematical models of real physical processes.

10.2.2 Classical Dynamics

We apply Hamiltonian dynamics on the price phase space. As in classical mechanics for material objects, we introduce a new variable $p = mv$, the *price momentum* variable. Instead of the price change vector $v = (v_1, \ldots, v_n)$, we consider the price momentum vector $p = (p_1, \ldots, p_n)$, $p_j = m_j v_j$. The space of price momenta is denoted by the symbol P. The space $\Omega = Q \times P$ will also be called the price phase space. *Hamiltonian equations* of motion on the price phase space have the form $\dot{q} = \partial H / \partial p_j$, $\dot{p}_j = -\partial H / \partial q_j$, $j = 1, \ldots, n$.

If the financial energy has the form (10.1) then the Hamiltonian equations have the form

$$\dot{q}_j = \frac{p_j}{m_j} = v_j, \quad \dot{p}_j = -\frac{\partial V}{\partial q_j}.$$

The latter equation can be written as

$$m_j \dot{v}_j = -\frac{\partial V}{\partial q_j}.$$

It is natural to call the quantity

$$\dot{v}_j = \lim_{\Delta t \to 0} \frac{v_j(t + \Delta t) - v_j(t)}{\Delta t}$$

the *price acceleration* (rate of change of price rate of change). The quantity

$$f_j(q) = -\frac{\partial V}{\partial q_j}$$

is called the (potential) financial force. We get the financial variant of Newton's second law:

$$m\dot{v} = f \tag{10.2}$$

Law 10.1. The product of the financial mass and the price acceleration is equal to the financial force.

In fact, the Hamiltonian evolution is determined by the following fundamental property of the financial energy: *The financial energy is not changed in the process of Hamiltonian evolution:*

$$H(q_1(t), \ldots, q_n(t), p_1(t), \ldots, p_n(t)) = H(q_1(0), \ldots q_n(0), p_1(0), \ldots, p_n(0)).$$

We need not restrict our considerations to financial energies of form (10.1). First of all external (e.g. economic) conditions as well as the character of interactions between traders at the market depend strongly on time. This must be taken into account by considering time-dependent potentials:

$$V = V(t, q).$$

Moreover, the assumption that the financial potential depends only on prices, $V = V(t, q)$, is not so natural for the modern financial market. Financial agents have complete information on price changes. This information is taken into account by traders for acts of arbitrage, see [237] for details. Therefore, it can be useful to consider potentials that depend not only on prices, but also on price changes: $V = V(t, q, v)$, or in the Hamiltonian framework: $V = V(t, q, p)$. In such a case the financial force is not potential. Therefore, it is also useful to consider the financial Newton's second law for general financial forces: $m\dot{v} = f(t, q, p)$.

Remark 10.1 (On the form of the kinetic financial energy) We copied the form of kinetic energy from classical mechanics for material objects. It may be that such a form of kinetic financial energy is not justified by a real financial market. It might be better to consider our choice of the kinetic financial energy as just the basis for mathematical modeling (and look for other possibilities).

Remark 10.2 (Domain of price dynamics) It is natural to consider a model in which all prices are nonnegative, $q_j(t) \geq 0$. Therefore financial Hamiltonian dynamics should be considered in the phase space $\Omega_+ = \mathbf{R}_+^n \times \mathbf{R}^n$, where $\mathbf{R}_+$ is the set of nonnegative real numbers. We shall not study this problem in detail, because our aim is the study of the corresponding quantum dynamics, but in the quantum case this problem is solved easily. One should just consider the corresponding Hamiltonian in the space of square integrable functions $L_2(\Omega_+)$. Another possibility in the classical case is to consider centered dynamics of prices: $z_j(t) = q_j(t) - q(0)$. The centered price $z_j(t)$ evolves in the configuration space $\mathbf{R}^n$.

10.2.3 Critique of Classical Econophysics

The model of Hamiltonian price dynamics on the price phase space can be useful to describe a market that depends only on "hard" economic conditions: natural resources, volumes of production, human resources and so on. However, the classical price dynamics cannot be applied (at least directly) to modern financial markets. It is clear that the stock market is not based only on these "hard" factors. There are other factors, soft ones (behavioral), that play an important (and sometimes even determining) role in forming of prices at the financial market. Market psychology should be taken into account. Negligibly small amounts of information (due to the rapid exchange of information) imply large changes of prices at the financial market. We can consider a model in which financial (psychological) waves are permanently present at the market. Sometimes these waves produce uncontrollable changes of prices disturbing the whole market (financial crashes). Of course, financial waves also depend on "hard economic factors." However, these factors do not play a crucial role in the formation of financial waves. Financial waves are merely waves of information.

We could compare the behavior of a financial market with the behavior of a gigantic ship that is ruled by a radio signal. A radio signal with negligibly small physical energy can essentially change (due to information contained in this signal) the motion of the gigantic ship. If we do not pay attention to (do not know about the presence of) the radio signal, then we will be continuously surprised by the ship's behavior. It can change its direction of motion without any "hard" reason (weather, destination, technical state of ship's equipment). However, if we know about the existence of radio monitoring, then we could find information that is sent by radio. This would give us a powerful tool to predict the ship's trajectory. This example on ship's monitoring was taken from the book by Bohm and Hiley [40] on so-called pilot wave quantum theory (or Bohmian quantum mechanics).

10.3 Quantum-like Econophysical Model of the Financial Market

10.3.1 Financial Pilot Waves

If we interpret the pilot wave as a field, then it is a rather strange field. It differs crucially from "ordinary physical fields," i.e., the electromagnetic field. We mention some of the pathological features of the pilot wave field. In particular, the force induced by this pilot wave field does not depend on the amplitude of the wave. Thus small waves and large waves equally disturb the trajectory of an elementary particle. Such features of the pilot wave make it possible to speculate [40] that this is just a wave of information (active information). Hence, the pilot wave field describes the propagation of information. The pilot wave is more similar to a radio signal that guides a ship. Of course, this is just an analogy (because a radio signal is related to an ordinary physical field, namely, the electromagnetic field). The more precise analogy is to compare the pilot wave with information contained in the radio signal.

We remark that the pilot wave (Bohmian) interpretation of quantum mechanics is not the conventional one. A few critical arguments against Bohmian quantum formalism can be mentioned:

1. Bohmian theory makes it possible to provide a mathematical description of the trajectory $q(t)$ of an elementary particle. However, such a trajectory does not exist according to conventional quantum formalism.
2. Bohmian theory is not local, namely, by means of the pilot wave field one particle "feels" another at large distances.

We say that these disadvantages of the theory will become advantages in our applications of Bohmian theory to the financial market. We also recall that Bohm and Hiley [40] and Hiley and Pilkkänen[144] have already discussed the possibility of interpreting the pilot wave field as a kind of information field. This information interpretation was essentially developed in my work, see, e.g., [176] devoted to pilot wave cognitive models.

Our fundamental assumption is that agents of the modern financial market are not just "classical-like agents." Their actions are ruled not only by classical-like financial potentials $V(t, q_1, \ldots, q_n)$, but also (in the same way as in the pilot wave theory for quantum systems) by an additional information (or psychological) potential induced by a financial pilot wave.

Therefore we cannot use classical financial dynamics (Hamiltonian formalism) on the financial phase space to describe the real price trajectories. Information (psychological) perturbation of Hamiltonian equations for price and price change must be taken into account. To describe such a model mathematically, it is convenient to use an object such as a *financial pilot wave* that rules the financial market.

In some sense $\psi(t, q)$ describes the psychological influence of the price configuration q on the behavior of agents of the financial market. In particular, $\psi(t, q)$ contains the expectations of agents.

We point out two important features of the financial pilot wave model:

1. All shares are coupled on the information level. The general formalism of the pilot wave theory says that if the function $\psi(t, q_1, \ldots, q_n)$ is not factorized, i.e.,

$$\psi(t, q_1, \ldots, q_n) \neq \psi_1(t, q_1) \ldots \psi_n(t, q_n),$$

then any change in the price q_i will automatically change the behavior of all agents of the financial market (even those who have no direct coupling with i-shares). This will imply a change in prices of j-shares for $i \neq j$. At the same time the "hard" economic potential $V(q_1, \ldots, q_n)$ need not contain any interaction term.

For example, let us consider for the moment the potential $V(q_1, \ldots, q_n) = q_1^2 + \ldots + q_n^2$. The Hamiltonian equations for this potential – in the absence of the financial pilot wave – have the form: $\dot{q}_j = p_j, \dot{p}_j = -2q_j, j = 1, 2, \ldots, n$. Thus the classical price trajectory $q_j(t)$, does not depend on the dynamics of prices of shares for other traders $i \neq j$ (for example, the price of Ericsson shares does not depend on the price of Nokia shares and vice versa).[5]

However, if, for example, the wave function has the form

$$\psi(q_1, \ldots, q_n) = ce^{i(q_1 q_2 + \ldots + q_{n-1} q_n)} e^{-(q_1^2 + \ldots + q_n^2)},$$

where $c \in C$ is some normalization constant, then financial behavior of agents of the financial market is nonlocal (see further considerations).

2. Reactions of the market do not depend on the amplitude of the financial pilot wave: waves $\psi, 2\psi, 100000\psi$ will produce the same reaction. Such a behavior of the market is quite natural (if the financial pilot wave is interpreted as an information wave, the wave of financial information). The amplitude of an information signal does not play so large a role in information exchange. Most important is the context of such a signal. The context is given by the shape of the signal, the form of the financial pilot wave function.

10.3.2 Dynamics of Prices Guided by Financial Pilot Wave

In fact, we do not need to develop a new mathematical formalism. We will just apply the standard pilot wave formalism to the financial market. The fundamental postulate of the pilot wave theory is that the pilot wave (field)

$$\psi(t, q_1, \ldots, q_n)$$

---

[5] Such a dynamics would be natural if these corporations operated on independent markets, e.g., Ericsson in Sweden and Nokia in Finland. Prices of their shares would depend only on local market conditions, e.g., on capacities of markets or consumer activity.

induces a new (quantum) potential

$$U(t, q_1, \ldots, q_n)$$

which perturbs the classical equations of motion. A modified Newton equation has the form

$$\dot{p} = f + g, \tag{10.3}$$

where $f = -\partial V / \partial q$ and $g = -\partial U / \partial q$. We call the additional financial force g a *financial mental force*. This force $g(t, q_1, \ldots, q_n)$ determines a kind of collective consciousness of the financial market. Of course, g depends on economic and other "hard" conditions given by the financial potential $V(t, q_1, \ldots, q_n)$. However, this is not a direct dependence. In principle, a nonzero financial mental force can be induced by the financial pilot wave ψ in the case of zero financial potential, $V \equiv 0$. So $V \equiv 0$ does not imply that $U \equiv 0$. *Market psychology is not totally determined by economic factors.* Financial (psychological) waves of information need not be generated by changes in the real economic situation. They are mixtures of mental and economic waves. Even in the absence of economic waves, mental financial waves can have a large influence on the market.

By using the standard pilot wave formalism we obtain the following rule for computing the financial mental force. We represent the financial pilot wave $\psi(t, q)$ in the form

$$\psi(t, q) = R(t, q)e^{iS(t,q)},$$

where $R(t, q) = |\psi(t, q)|$ is the amplitude of $\psi(t, q)$ (the absolute value of the complex number $c = \psi(t, q)$) and $S(t, q)$ is the phase of $\psi(t, q)$ (the argument of the complex number $c = \psi(t, q)$). Then the financial mental potential is computed as

$$U(t, q_1, \ldots, q_n) = -\frac{1}{R} \sum_{i=1}^{n} \frac{\partial^2 R}{\partial q_i^2}(t, q_1, \ldots, q_n)$$

and the financial mental force as

$$g_j(t, q_1, \ldots, q_n) = \frac{-\partial U}{\partial q_j}(t, q_1, \ldots, q_n).$$

These formulas imply that strong financial effects are produced by financial waves having significant variations of amplitude.

Example 10.1 (Financial waves with small variation have no effect) Let us start with the simplest example: $R \equiv const$. Then the financial (behavioral) force $g \equiv 0$. As $R \equiv const$, it is impossible to change expectations of the whole financial market by varying the price q_j of one fixed type of shares, j. The constant information field does not induce psychological financial effects at all. As we have already remarked, the absolute value of this constant does not play any role. Waves of constant amplitude $R = 1$, as well as $R = 10^{100}$, produce no effect.

Let now consider the case $R(q) = cq$, $c > 0$. This is a linear function; variation is not so large. As a result, $g \equiv 0$ here also. There are no financial behavioral effects.

Example 10.2 (Speculation) Let $R(q) = c(q^2 + d)$, $c, d > 0$. Here

$$U(q) = -\frac{2}{q^2 + d}$$

(it does not depend on the amplitude c !) and

$$g(q) = \frac{-4q}{(q^2 + d)^2}.$$

The quadratic function varies essentially more strongly than the linear function, and, as a result, such a financial pilot wave induces a nontrivial financial force.

We analyze financial drives induced by such a force. We consider the following situation: (the starting price) $q > 0$ and $g < 0$. The financial force g stimulates the market (which works as a huge cognitive system) to decrease the price. For small prices, $g(q) \approx -4q/d^2$. If the financial market increases the price q for shares of this type, then the negative reaction of the financial force becomes stronger and stronger. The market is pressed (by the financial force) to stop increasing the price q. However, for large prices, $g(q) \approx -4/q^3$. If the market can approach this range of prices (despite the negative pressure of the financial force for relatively small q) then the market will feel a decrease of the negative pressure (we recall that we consider the financial market as a huge cognitive system). This model explains well the speculative behavior of the financial market.

Example 10.3 Let now $R(q) = c(q^4 + b)$, $c, b > 0$. Thus

$$g(q) = \frac{bq - q^5}{(q^4 + b)^2}.$$

Here the behavior of the market is more complicated. Set $d =^4 \sqrt{b}$. If the price q is changing from $q = 0$ to $q = d$ then the market is motivated (by the financial force $g(q)$) to increase the price. The price $q = d$ is critical for its financial activity. For psychological reasons (of course, indirectly based on the whole information available at the market) the market "understands" that it would be dangerous to continue to increase the price. After approaching the price $q = d$, the market has a psychological stimulus to decrease the price.

Financial pilot waves $\psi(q)$ with $R(q)$ that are polynomials of higher order can induce very complex behavior. The interval $[0, \infty)$ is split into a collection of subintervals $0 < d_1 < d_2 < \ldots < d_n < \infty$ such that at each price level $q = d_j$ the trader changes his attitude to increase or to decrease the price.

In fact, we have considered just a one-dimensional model. In the real case we have to consider multidimensional models of huge dimension. A financial pilot wave $\psi(q_1, \ldots, q_n)$ on such a price space Q induces splitting of Q into a large number of domains $Q = O_1 \bigcup \ldots \bigcup O_N$.

The only problem that we have still to solve is the description of the time-dynamics of the financial pilot wave, $\psi(t, q)$. We follow the standard pilot wave theory. Here $\psi(t, q)$ is found as the solution of Schrödinger's equation. The Schrödinger equation for the energy

$$H(q, p) = \frac{1}{2} \sum_{j=1}^{n} \frac{p_j^2}{m_j} + V(q_1, \ldots, q_n)$$

has the form

$$i\hbar \frac{\partial \psi}{\partial t}(t, q_1, \ldots, q_n) =$$
$$-\sum_{j=1}^{n} \frac{\hbar^2}{2m_j} \frac{\partial^2 \psi(t, q_1, \ldots, q_n)}{\partial q_j^2} + V(q_1, \ldots, q_n)\psi(t, q_1, \ldots, q_n), \qquad (10.4)$$

with the initial condition

$$\psi(0, q_1, \ldots, q_n) = \psi(q_1, \ldots, q_n).$$

Thus if we know $\psi(0, q)$ then by using Schrödinger's equation we can find the pilot wave at any instant of time t, $\psi(t, q)$. Then we compute the corresponding mental potential $U(t, q)$ and mental force $g(t, q)$ and solve Newton's equation.

We shall use the same equation to find the evolution of the financial pilot wave. We have only to make one remark, namely, on the role of the constant $\hbar$ in Schrödinger's equation, see [132–136]. In quantum mechanics (which deals with microscopic objects) $\hbar$ is the Dirac constant, which is based on the Planck constant h. The latter constant plays the fundamental role in all quantum considerations. However, originally h appeared as just a scaling numerical parameter for processes of energy exchange. Therefore in our financial model we can consider $\hbar$ as a price scaling parameter, namely, the unit in which we would like to measure price change. We do not present any special value for $\hbar$. There are numerous investigations into price scaling. It may be that there can be recommended some special value for $\hbar$ related to the modern financial market, a *fundamental financial constant*. However, it seems that

$$\hbar = \hbar(t)$$

evolves depending on economic development.

We suppose that the financial pilot wave evolves according to the financial Schrödinger equation (an analogue of Schrödinger's equation) on the price space.

In the general case this equation has the form

$$i\hbar \frac{\partial \psi}{\partial t}(t, q) = \widehat{H} \psi(t, q), \quad \psi(0, q) = \psi(q),$$

where $\widehat{H}$ is a self-adjoint operator corresponding to the financial energy given by a function $H(q, p)$ on the financial phase space. Here we proceed in the same way as in ordinary quantum theory for elementary particles.

10.4 Application of Quantum Formalism to the Financial Market

We now turn back to the general scheme, concentrating on the configuration representation, $\psi : Q \to \mathbf{C}$; $\psi \in L_2(Q) \equiv L_2(Q, dx)$. This is the general QL statistical formalism on the price space.

As in ordinary quantum mechanics, we consider a representation of financial quantities, observables, by symmetric operators in $L_2(Q)$. By using Schrödinger's representation we define price and price change operators by setting

$$\hat{q}_j \psi(q) = q_j \psi(q),$$

the operator of multiplication by the q_j-price;

$$\hat{p}_j = \frac{\hbar}{i} \frac{\partial}{\partial q_j},$$

the operator of differentiation with respect to the q_j-price, normalized by the scaling constant $\hbar$. Operators of price and price change satisfy the canonical commutation relations

$$[\hat{q}, \hat{p}] = \hat{q}\hat{p} - \hat{p}\hat{q} = i\hbar.$$

By using this operator representation of price and price changes we can represent every function $H(q, p)$ on the financial phase space as an operator $H(\hat{q}, \hat{p})$ in $L_2(Q)$. In particular, the financial energy operator is represented by the operator

$$\widehat{H} = \sum_{j=1}^{n} \frac{\hat{p}_j^2}{2m_j} + V(\hat{q}_1, \ldots, \hat{q}_n) = -\sum_{j=1}^{n} \frac{\hbar^2}{2m_j} \frac{\partial^2}{\partial q_j^2} + V(q_1, \ldots, q_n).$$

Here $V(\hat{q}_1, \ldots, \hat{q}_n)$ is the operator of multiplication by the function $V(q_1, \ldots, q_n)$.

In this general QL formalism for the financial market we do not consider individual evolution of prices (in contrast to the Bohmian approach). The theory is purely statistical. We can only determine the average of a financial observable A for some fixed state ϕ of the financial market:

$$\langle A \rangle_\phi = \langle A\phi, \phi \rangle.$$

The use of the Bohmian model gives the additional possibility of determining individual trajectories.

10.5 Standard Deviation of Price

We are interested in the standard deviation of the price q_t. Let ψ be the mental state of the financial market. The quantum formalism gives us the following formula for the price dispersion:

$$\sigma_\psi^2(q_t) = E_\psi q_t^2 - (E_\psi q_t)^2, \tag{10.5}$$

where for an observable a the quantum average (with respect to the state ψ) is given by $E_\psi a = \langle a\psi, \psi \rangle$.

Since, for any observable a_t,

$$E_\psi a_t = E_{\psi(t)} a_0, \tag{10.6}$$

we have

$$\sigma_\psi^2(q_t) = E_{\psi(t)} q^2 - (E_{\psi(t)} q)^2. \tag{10.7}$$

So

$$\sigma_\psi^2(q_t) = \langle q^2 \psi(t), \psi(t) \rangle - \langle q\psi(t), \psi(t) \rangle^2. \tag{10.8}$$

Suppose that at the initial instant of time the wave function has the form of a Gaussian packet:

$$\psi_0(q) \approx \int_{-\infty}^{+\infty} \exp\{-k^2(\Delta q)^2 + ikq\}dk,$$

where Δq is the width of packet in the price space. Here the mean value of price is equal to zero. It is well known that

$$\psi(t, q) \approx \int_{-\infty}^{+\infty} \exp\{-k^2(\Delta q)^2 + ikq - (ihk^2 t)/2m\}dk.$$

Here the mean value of price is equal to zero for any instance of time. By calculating this integral we see that

$$\sigma_\psi(q_t) = \sqrt{\langle q^2 \psi(t), \psi(t) \rangle - \langle q\psi(t), \psi(t) \rangle^2} = \sqrt{\langle q^2 \psi(t), \psi(t) \rangle} \approx ht/m\Delta q$$

for large t.

Thus for a Gaussian packet of prices its standard deviation evolves as a linear function with respect to t. Large financial mass (i.e., a higher level of emission of shares) induces smaller standard deviation – so the price does not fluctuate far from the mean value. If the level of emission is very small, then large deviations from the mean value can be expected.

10.6 Comparison with Conventional Models of the Financial Market

Our model of the stocks market differs crucially from the main conventional models. Therefore we should perform an extended comparative analysis of our model and known models. This is not a simple task and it takes a lot of effort.

10.6.1 Stochastic Model

Since the pioneer paper of Bachelier [24], various models of the financial market based on stochastic processes have been actively developed. We recall that Bachelier determined the probability of price changes $P(v(t) \leq v)$ by writing down what is now called the Chapman–Kolmogorov equation. If we introduce the density of this probability distribution $p(t, x)$, so $P(x_t \leq x) = \int_{-\infty}^{x} p(t, x)dx$, then it satisfies the Cauchy problem of the partial differential equation of the second order. This equation is known in physics as Chapman's equation and in probability theory as the direct Kolmogorov equation. In the simplest case, when the underlying diffusion process is the Wiener process (Brownian motion), this equation has the form (the heat conduction equation)

$$\frac{\partial p(t, x)}{\partial t} = \frac{1}{2} \frac{\partial^2 p(t, x)}{\partial x^2}. \tag{10.9}$$

We recall again that in Bachelier's paper [24], $x = v$ was the price change variable.
    For a general diffusion process we have the direct Kolmogorov equation

$$\frac{\partial p(t, x)}{\partial t} = \frac{1}{2} \frac{\partial^2}{\partial x^2} (\sigma^2(t, x)p(t, x)) - \frac{\partial}{\partial x} (\mu(t, x)p(t, x)). \tag{10.10}$$

This equation is based on the diffusion process

$$dx_t = \mu(t, x_t) dt + \sigma(t, x_t) dw_t, \tag{10.11}$$

where $w(t)$ is the Wiener process. This equation should be interpreted as a slightly colloquial way of expressing the corresponding integral equation

$$x_t = x_{t_0} + \int_{t_0}^{t} \mu(s, x_s)ds + \int_{t_0}^{t} \sigma(s, x_s)dw_s. \qquad (10.12)$$

We remark that Bachelier's original proposal of Gaussian-distributed price changes was soon replaced by a model in which prices of stocks are *log–normal distributed*, i.e., stock prices $q(t)$ are performing a *geometric Brownian motion*. In a geometric Brownian motion, the difference of the logarithms of prices are Gaussian distributed.

We recall that a stochastic process S_t is said to follow a geometric Brownian motion if it satisfies the following stochastic differential equation:

$$dS_t = u\, S_t\, dt + v\, S\, dw_t, \qquad (10.13)$$

where w_t is a Wiener process (Brownian motion) and u ("the percentage drift") and v ("the percentage volatility") are constants. The equation has an analytic solution:

$$S_t = S_0 \exp\left((u - v^2/2)t + vw_t\right). \qquad (10.14)$$

The $S_t = S_t(\omega)$ depends on a random parameter ω; this parameter is typically omitted. The crucial property of the stochastic process S_t is that the random variable

$$\log(S_t/S_0) = \log(S_t) - \log(S_0)$$

is normally distributed.

In contrast to such stochastic models, our Bohmian model of the stock market is not based on the theory of stochastic differential equations. In our model the randomness of the stock market cannot be represented in the form of some transformation of the Wiener process.

We recall that the stochastic process model has been intensely criticized for many reasons, see, e.g., [237]. First of all there are a number of difficult problems that could be interpreted as technical problems. The most important among them is the problem of the choice of an adequate stochastic process $\xi(t)$ describing price or price change. Nowadays it is widely accepted that the geometric Bohmian motion model provides only a first approximation of what is observed in real data. One should try to find new classes of stochastic processes. In particular, they would provide an explanation of the empirical evidence that the tails of measured distributions are longer than expected for a geometric Brownian motion. To solve this problem, Mandelbrot proposed that the price changes should be considered to follow a *Levy distribution* [237]. However, the Levy distribution has a rather pathological property: its variance is infinite. Therefore, as was emphasized in the book by Mantegna and Stanley [237], the problem of finding a stochastic process providing an adequate description of the stock market is still unsolved.

However, our critique of the conventional stochastic processes approach to the stock market has no direct relation to this discussion on the choice of an underlying stochastic process. We are closer to scientific groups that criticize this conventional model by questioning the possibility of describing price dynamics by stochastic processes at all.

10.6.2 Deterministic Dynamical Model

In particular, a lot of work has been done on applying deterministic nonlinear dynamical systems to simulate financial time series, see [237] for details. This approach is typically criticized through the following general argument: "the time evolution of an asset price depends on all information affecting the investigated asset and it seems unlikely to us that all this information can be essentially described by a small number of nonlinear equations," [237]. We support such a viewpoint.

We shall use only critical arguments against the hypothesis of the stochastic stock market that were provided by adherents of the hypothesis of a deterministic (but essentially nonlinear) stock market.

Only at first sight is the Bohmian financial model a kind of deterministic model. Of course, dynamics of prices (as well as price changes) are deterministic. It is described by Newton's second law, see the ordinary differential equation (10.3). It seems that randomness can be incorporated into such a model only through the initial conditions:

$$\dot{p}(t, \omega) = f(t, q(t, \omega)) + g(t, q(t, \omega)), \quad q(0) = q_0(\omega), \quad p(0) = p_0(\omega), \quad (10.15)$$

where $q(0) = q_0(\omega)$, $p(0) = p_0(\omega)$ are random variables (initial distribution of prices and momenta) and ω is a chance parameter.

However, the situation is not so simple. Bohmian randomness does not reduce to randomness of initial conditions or chaotic behavior of (10.3) for some nonlinear classical and quantum forces. These are classical impacts on randomness. But a really new impact is given by the essentially quantum randomness that is encoded in the ψ-function (i.e., pilot wave or wave function). As we know, the evolution of the ψ-function is described by an additional equation – Schrödinger's equation – and hence the ψ-randomness can be extracted neither from the initial conditions for (10.15) nor from possible chaotic behavior.

In our model the ψ-function gives the dynamics of expectations at the financial market. These expectations are a huge source of randomness at the market – mental (psychological) randomness. However, this randomness is not classical (so it is a non-Kolmogorov probability model).

Finally, we remark that in quantum mechanics the wave function is not a measurable quantity. It seems that we have a similar situation for the financial market. We are not able to measure the financial ψ-field (which is an infinite-dimensional object, since the Hilbert space has infinite dimension). This field contains thoughts and expectations of millions of agents and of course it could not be "recorded" (unlike prices or price changes).

10.6.3 Stochastic Model and Expectations of Agents of the Financial Market

Let us consider again the model of the stock market based on geometric Brownian motion:

$$dS_t = uS_t \, dt + vS \, dw_t.$$

We notice that in this equation there is no term describing the behavior of agents of the market. Coefficients u and v do not have any direct relation to expectations and the market psychology. Moreover, even if we introduce some additional stochastic processes

$$\eta(t, \omega) = (\eta_1(t, \omega), \ldots, \eta_N(t, \omega))$$

describing the behavior of agents and additional coefficients (in stochastic differential equations for such processes) we would not be able to simulate the real market. A finite-dimensional vector $\eta(t, \omega)$ cannot describe the "mental state of the market", which is of infinite complexity. One can consider the Bohmian model as the introduction of the infinite-dimensional chance parameter ψ. And this chance parameter cannot be described by classical probability theory.

Chapter 11
The Problem of Smoothness of Bohmian Trajectories

We point out that there are two basic (rather different) interpretations of Bohmian mechanics: *the quantum force interpretation*, Bohm and Hiley [40], and *the guidance-field interpretation*, e.g., Cushing et al. [72]. In the first, the basic equation is Newton's equation for the position of a quantum particle, and in the second, the guidance equation for its momentum.

One objection (presented by Dr. Roger Pettersson at the University of Växjö) to applying the Bohmian quantum formalism to describe the dynamics of prices (of e.g. shares) of the financial market is the smoothness of trajectories obtained in the Bohmian model – at least if one uses the *quantum force interpretation,* Bohm and Hiley [40]. In contrast to this, in financial mathematics it is commonly assumed that price trajectories are not differentiable, Mantegna and Stanley [237] or Shiryaev [281]. This is a problem.

This problem can be easily solved by using the guidance-field interpretation of Bohmian mechanics (i.e. if one proceeds without introducing the quantum-like financial force, Cushing et al. [72]) and considering the integral version of the guidance equation. However, another objection can be presented even in this case (and it was really presented by Roger Pettersson). Solutions of the integral guidance equation have zero quadratic variation. In contrast to this, in financial mathematics it is commonly assumed that price trajectories have nonzero quadratic variations, Mantegna and Stanley [237] or Shiryaev [281]. This is also a problem.

It seems that independently of the interpretation of Bohmian mechanics one cannot apply it to the financial market in the canonical deterministic form. The only way to proceed with real financial data is to apply the stochastic version of the pilot wave theory – the model of Bohm–Vigier [40].

11.1 Existence Theorems for Nonsmooth Financial Forces

11.1.1 The Problem of Smoothness of Price Trajectories

In the Bohmian model for price dynamics the price trajectory $q(t)$ can be found as the solution of the equation

A. Khrennikov, *Ubiquitous Quantum Structure*,
DOI 10.1007/978-3-642-05101-2_11, © Springer-Verlag Berlin Heidelberg 2010

$$m\frac{d^2q(t)}{dt^2} = f(t, q(t)) + g(t, q(t)) \qquad (11.1)$$

with the initial condition $q(t_0) = q_0, q'(t_0) = q_0'$. Here we consider a "classical" (time-dependent) force

$$f(t, q) = -\frac{\partial V(t, q)}{\partial q}$$

and a "quantum-like" force

$$g(t, q) = -\frac{\partial U(t, q)}{\partial q},$$

where $U(t, q)$ is the quantum potential, induced by the Schrödinger dynamics. In Bohmian mechanics for *physical systems*, (11.1) is considered as an ordinary differential equation and $q(t)$ as the unique solution (corresponding to the initial conditions $q(t_0) = q_0, q'(t_0) = q_0'$) of the class C^2 : $q(t)$ is assumed to be twice differentiable with continuous $q''(t)$.

One possible objection to applying the Bohmian quantum model to describe dynamics of prices (of e.g. shares) at the financial market is smoothness of trajectories. As mentioned above, in financial mathematics it is commonly assumed that the price-trajectory is not differentiable, see, e.g., [237, 281].

Of course, one could simply reply that there are no smooth trajectories in nature. Smooth trajectories belong neither to physical nor financial reality. They appear in mathematical models that can be used to describe reality. It is clear that the possibility of applying a mathematical model with smooth trajectories depends on the chosen time scale. Trajectories that can be considered as smooth (or continuous) at one time scale might be nonsmooth (or discontinuous) at a finer time scale.

We illustrate this general philosophic thesis by the history of development of financial models. We recall that at the first stage of development of financial mathematics, in the Bachelier model and the Black and Scholes model, *processes with continuous trajectories were considered*: the Wiener process and more general diffusion processes. However, recently it was claimed that such stochastic models (with continuous processes) are not completely adequate for real financial data, see, e.g., [237, 281] for detailed analysis. It was observed that at finer time scales some Levy processes with jump trajectories are more adequate for data from the financial market.

Therefore one could say that the Bohmian model provides a rough description of price dynamics and describes not the real price trajectories but their smoothed versions. However, it would be interesting to keep the interpretation of Bohmian trajectories as the real price trajectories. In such an approach one should obtain nonsmooth Bohmian trajectories. The following section is devoted to theorems providing nonsmooth solutions.

11.1.2 Picard's Theorem and its Generalization

We recall the standard uniqueness and existence theorem for ordinary differential equations, Picard's theorem, that guarantees smoothness of trajectories, see, e.g., [220].

Theorem 11.1 *Let $F : [0, T] \times \mathbf{R} \to \mathbf{R}$ be a continuous function and let F satisfy the Lipschitz condition with respect to the variable x:*

$$|F(t, x) - F(t, y)| \leq c|x - y|, \quad c > 0. \tag{11.2}$$

Then, for any point $(t_0, x_0) \in [0, T) \times \mathbf{R}$ there exists the unique C^1-solution of the Cauchy problem:

$$\frac{dx}{dt} = F(t, x(t)), \quad x(t_0) = x_0, \tag{11.3}$$

on the segment $\Delta = [t_0, a]$, where $a > 0$ depends on t_0, x_0, and F.

We recall the standard proof of this theorem, because the scheme of this proof can be easily generalized to prove Theorems 11.3 and 11.4. Let us consider the space of continuous functions $x : [t_0, a] \to \mathbf{R}$, where $a > 0$ is a number which will be determined. Denote this space by the symbol $C[t_0, a]$. The Cauchy problem (11.3) for the ordinary differential equation can be written as the integral equation

$$x(t) = x(t_0) + \int_{t_0}^{t} F(s, x(s))ds \tag{11.4}$$

The crucial point for our further considerations is that continuity of the function F with respect to the pair of variables (t, x) implies continuity of $y(s) = F(s, x(s))$ for any continuous $x(s)$. But the integral $z(t) = \int_0^t y(s)ds$ is differentiable for any continuous $y(s)$ and $z'(t) = y(t)$ is also continuous. The basic point of the standard proof is that, for a sufficiently small $a > 0$, the operator

$$G(x)(t) = x_0 + \int_{t_0}^{t} F(s, x(s))ds \tag{11.5}$$

maps the functional space $C[t_0, a]$ into $C[t_0, a]$ and there is a contraction in this space:

$$\rho_\infty(G(x_1), G(x_2)) \leq \alpha\rho_\infty(x_{10}, x_{20}), \quad \alpha < 1, \tag{11.6}$$

for any two trajectories $x_1(t), x_2(t) \in C[t_0, a]$ such that $x_1(t_0) = x_{10}$ and $x_2(t_0) = x_{20}$. Here, to obtain $\alpha < 1$, the interval $[t_0, a]$ should be chosen sufficiently small, see further considerations. Here $\rho_\infty(u_1, u_2) = ||u_1 - u_2||_\infty$ and $||u||_\infty = \sup_{t_0 \le t \le a} |u(s)|$. The contraction condition, $\alpha < 1$, implies that the iterations

$$x_1(t) = x_0 + \int_{t_0}^t F(S, x_0)ds,$$

$$x_2(t) = x_0 + \int_{t_0}^t F(S, x_1(S))ds, ...,$$

$$x_n(t) = x_0 + \int_{t_0}^t F(S, x_{n-1}(S))ds, ...$$

converge to a solution $x(t)$ of the integral equation (11.4). Finally, we remark that the contraction condition (11.6) implies that the solution is unique in the space $C[t_0, a]$.

We also recall the well known Peano theorem, [220]:

Theorem 11.2 *Let* $F : [0, T] \times \mathbf{R}$ *be a continuous function. Then, for any point* $(t_0, x_0) \in [0, T] \times \mathbf{R}$ *there exists locally a* C^1-*solution of the Cauchy problem (11.3).*

We remark that Peano's theorem does not imply uniqueness of solution.

It is clear that discontinuous financial forces can induce price trajectories $q(t)$ that are not smooth: moreover, price trajectories can even be discontinuous! From this point of view the main problem is not smoothness of price trajectories $q(t)$ (and in particular the zero covariation for such trajectories), but the absence of an existence and uniqueness theorem for discontinuous financial forces. We shall formulate and prove such a theorem. Of course, outside the class of smooth solutions one could not study the original Cauchy problem for an ordinary differential equation (11.3). Instead of this one should consider the integral equation (11.4).

We shall generalize Theorem 11.1 to discontinuous F. Let us consider the space $BM[t_0, a]$ consisting of bounded measurable functions $x : [t_0, a] \to \mathbf{R}$. Thus: a) $\sup_{t_0 \le t \le a} |x(t)| \equiv ||x||_\infty < \infty$; b) for any Borel subset $A \subset \mathbf{R}$, its preimage $x^{-1}(A) = \{s \in [t_0, a] : x(s) \in A\}$ is again a Borel subset in $[t_0, a]$.

Lemma 11.1 *The space of trajectories* $BM[t_0, a]$ *is a Banach space.*

Theorem 11.3 *Let* $F : [0, T] \times \mathbf{R} \to \mathbf{R}$ *be a measurable bounded function and let* F *satisfy the Lipschitz condition with respect to the* x-*variable, see (11.2). Then, for any point* $(t_0, x_0) \in [0, T) \times \mathbf{R}$, *there exists a unique solution of the integral equation (11.4) of the class* $BM[t_0, a]$, *where* $a > 0$ *depends on* x_0, t_0, *and* F.

Proposition 11.1 (Continuity of the solution of the integral equation) *Let the conditions of Theorem 11.3 hold. Then solutions are continuous functions* $x : [t_0, a] \to \mathbf{R}$.

Thus Theorem 11.3 gives a sufficient condition for the existence of the unique continuous trajectory solution $x(t)$ for the integral equation (11.4). But, of course, in general $x(t)$ is not continuously differentiable!

Theorem 11.4 *Let* f *satisfy the Lipschitz condition (11.2). Then for any point* $(t_0, x_0) \in [0, T) \times R)$ *there exists the unique solution of the integral equation (11.4) of the class* $L_2[t_0, a]$, *where* $a > 0$ *depends on* x_0, t_0, *and* F.

Proposition 11.2 (Continuity) *Let the conditions of Theorem 11.4 hold. Then solutions* $x : [t_0, a] \to \mathbf{R}$ *are continuous functions.*

Thus we again have obtained continuous, but in general non-smooth ($x \notin C^1$) solutions of the basic integral equation.

We remark that Theorems 11.3 and 11.4 are valid in the multidimensional case: $x_0 = (x_{01}, \ldots, x_{0n})$, $x(t) = (x_1(t), \ldots, x_n(t))$, and $F : [0, T] \times \mathbf{R}^n \to \mathbf{R}^n$.

To show this, we should change in all previous considerations the absolute value $|x|$ to be norm on the Euclidean space $\mathbf{R}^n$: $\|x\| = \sqrt{\sum_{j=1}^n x_j^2}$. We now use a standard trick to apply our theory to Newton's equation (11.1), which is a second-order differential equation. We rewrite this equation as a system of equations first order with respect to $x = (x_1, \ldots, x_n, x_{n+1}, x_{2n})$, where $x_1 = q_1, \ldots, x_n = q_n$, $x_{n+1} = p_1, \ldots, x_{2n} = p_n$. In fact, this is nothing other than the phase space representation. Newton's equation (11.1) will be written as the Hamilton equation. However, the Hamiltonian structure is not important for us in this context. In any event we obtain the following system of first-order equations:

$$\frac{dx}{dt} = F(t, x(t)), \tag{11.7}$$

where

$$F(t, x) = \begin{pmatrix} x_{n+1} \\ \cdot \\ \cdot \\ \cdot \\ x_{2n} \\ f_1(t, x_1, \ldots, x_n) + g_1(t, x_1, \ldots, x_n) \\ \cdot \\ \cdot \\ \cdot \\ f_n(t, x_1, \ldots, x_n) + g_n(t, x_1, \ldots, x_n) \end{pmatrix}.$$

Here $f_j(t, x_1, x_n) = \partial V / \partial x_j (t, x_1, \ldots, x_n)$ and $g_j(t, x_1, \ldots, x_n) = \partial U / \partial x_j$ $(t, x_1, \ldots, x_n)$. Therefore if

$$\nabla V = \left(\frac{\partial V}{\partial x_n}, \ldots, \frac{\partial V}{\partial x_n} \right)$$

or

$$\nabla U = \left(\frac{\partial U}{\partial x_n}, \ldots, \frac{\partial U}{\partial x_n} \right)$$

are not continuous, then the standard existence and uniqueness theorems, see Theorems 11.1 and 11.2 could not be applied. But, instead of the ordinary differential equation (11.7), we can consider the integral equation

$$x(t) = x_0 + \int_{t_0}^{t} F(s, x(s))ds \qquad (11.8)$$

and apply Theorems 11.3 and 11.4 to this equation. We note that owing to the structure of $F(t, x)$, we have in fact $p_1(t) = p_{01} + \int_{t_0}^{t} F_1(s, q(s))ds$, $p_n(t) = p_{0n} + \int_{t_0}^{t} F_n(s, q(s))ds$, $q_1(t) = q_{01} + 1/m \int_{t_0}^{t} p_1(s, q(s))ds$, $q_n(t) = q_{0n} + 1/m \int_{t_0}^{t} p_n(s)ds$. By Propositions 11.1 and 11.2 solutions $p_j(t)$ are continuous functions. Therefore integrals $\int_{t_0}^{t} p_j(s)ds$ are continuous differentiable functions. Thus under the conditions of Theorem 11.3 or Theorem 11.4 we obtain the following price dynamics:

Price trajectories are of the class C^1 (so $dq/dt(t)$ exists and is continuous), but price velocity $v(t) = p(t)/m$ is in general nondifferentiable.

11.2 The Problem of Quadratic Variation

The quadratic variation of a function u on an interval $[0, T]$ is defined as

$$\langle u \rangle(T) = \lim_{\|P\| \to 0} \sum_{k=0}^{n-1} (u(t_{k+1}) - u(t_k))^2,$$

where $P = \{0 = t_0 < t_1 < \ldots < t_n = T\}$ is a partition of $[0, T]$ and $\|P\| = \max_k\{(t_{k+1} - t_k)\}$. We recall the following well-known result:

Theorem 11.5 *If u is differentiable, then $\langle f \rangle(T) = 0$.*

Therefore, the quadratic variation of any smooth Bohmian trajectory is equal to zero. On the other hand, it is well known that real price trajectories have nonzero quadratic variation, [237, 281]. This is a strong objection to consideration of smooth Bohmian price trajectories.

In the previous section existence theorems were derived that provide nonsmooth trajectories. One might hope that solutions given by those theorems would have nonzero quadratic variation. But this is not the case.

Theorem 11.6 *Assume that $x(t) = x_0 + \int_0^t F(s, x(s))ds$, where F is bounded, i.e., $|F(t, x)| \leq K$, and measurable. Then the quadratic variation $\langle F \rangle(t) = 0$.*

Proof We have: $|x(t_k) - x(t_{k-1})|^2 = |\int_{t_{k-1}}^{t_k} F(s, x(s))ds|^2 \leq K^2(t_k - t_{k-1})^2$. Hence, with a partition of $[0, t]$, say, $0 = t_0 < t_1 < \ldots < t_n = t$, we get

$$\sum_{k=1}^{n} |x(t_k) - x(t_{k-1})|^2 \leq K^2 \sum_{1}^{n} (t_k - t_{k-1})^2$$

$$\leq K^2 \max_{k:1 \leq k \leq n} (t_k - t_{k-1}) \sum_{1}^{n} (t_k - t_{k-1}) = K^2 \max_{k:1 \leq k \leq n} (t_k - t_{k-1}),$$

which converges to zero as the partition gets finer, i.e. the quadratic variation of $t \mapsto x(t)$ is zero.

Thus the objection related to the nonzero quadratic variation is essentially stronger than the smoothness objection. One way to escape this problem is to consider unbounded quantum potentials or even potentials that are given by distributions.

11.3 Singular Potentials and Forces

We present some examples of discontinuous quantum forces g (induced by a discontinuous quantum potential U).

11.3.1 Example

Let us consider the wave function $\psi(x) = c(x + 1)^2 e^{-x^2/2} dx$, where c is the normalization constant providing $\int_{-\infty}^{+\infty} |\psi(x)|^2 dx = 1$. Here $\psi(x) \equiv R(x) = |\psi(x)|$. We have

$$R'(x) = c[2(x + 1) - x(x + 1)^2]e^{-\frac{x^2}{2}} = -c(x^3 + 2x^2 - x - 2)e^{-\frac{x^2}{2}},$$

and $R''(x) = c(x^4 + 2x^3 - 4x^2 - 6x + 1)e^{-x^2/2}$. Hence

$$U(x) = -\frac{R''(x)}{R(x)} = \frac{x^4 + 2x^3 - 4x^2 - 6x + 1}{(x + 1)^2}.$$

Thus the potential has a singularity at the point $x = -1$.

In this example a singularity in the quantum potential $U(t, x)$ is a consequence of division by the amplitude of the wave function $R(t, x)$. If $|\psi(t, x_0)| = 0$, then there can appear a singularity at the point x_0.

11.3.2 Singular Quantum Potentials

Let $\widehat{H}$ be a self-adjoint operator, $\widehat{H} \geq 0$, in $L_2(\mathbf{R}^n)$ (a Hamiltonian – an operator representing the financial energy). Let us consider the corresponding Schrödinger equation $\partial \psi / \partial t = \widehat{H} \psi$, $\psi(0) = \psi_0$, in $L_2(\mathbf{R}^n)$. Then its solution has the form

$$u_t(\psi_0) = e^{\frac{-it\widehat{H}}{h}} \psi_0.$$

If the operator $\widehat{H}$ is continuous, then its exponent is defined with aid of the usual exponential power series:

$$e^{\frac{-it\widehat{H}}{h}} = \sum_{n=0}^{\infty} \left(\frac{-it\widehat{H}}{h}\right)^n /n! = \sum_{n=0}^{\infty} \left(\frac{-it}{h}\right)^n /n! \, \widehat{H}^n.$$

If the operator $\widehat{H}$ is not continuous, then this exponent can be defined by using the *spectral theorem* for self-adjoint operators.

We recall that, for any $t \geq 0$, the map $u_t : L_2(\mathbf{R}^n) \rightarrow L_2(\mathbf{R}^n)$ is a unitary operator: (a) it is one-to-one; (b) it maps $L_2(\mathbf{R}^n)$ onto $L_2(\mathbf{R}^n)$; (c) it preserves the scalar product: $\langle u_t \psi, u_t \phi \rangle = \langle \psi, \phi \rangle$ $\psi, \phi \in L_2$.

We pay attention to (b). By (b), for any $\phi \in L_2(\mathbf{R}^n)$, we can find a $\psi_0 \in L_2(\mathbf{R}^n)$ such that $\phi = u_t(\psi_0)$. It is sufficient to choose $\psi_0 = u_t^{-1}(\phi)$ (any unitary operator is invertible). Thus, $\psi(t) = u_t(\psi_0) = \phi$. In general a function $\phi \in L_2(\mathbf{R}^n)$ is not a smooth or even continuous function! Therefore in the case under consideration (we created the wave function ψ such that $\psi(t) = \phi$, where ϕ was an arbitrarily chosen square integrable function),

$$U(t, x) = -\frac{|\psi(t, x)|''}{|\psi(t, x)|} = -\frac{|\phi(x)|''}{\phi(x)}$$

is in general a generalized function (distribution)! For example, let us choose

$$\phi(x) = \begin{cases} \dfrac{1}{2b}, & -b \leq x \leq b \\ 0, & x \notin [-b, b] \end{cases}$$

Here $R(t, x) = |\phi(x)| = \phi(x)$ and

$$R'(t, x) = \frac{\delta(x + b) - \delta(x - b)}{2b},$$

$$R''(t, x) = \frac{\delta'(x + b) - \delta'(x - b)}{2b}.$$

Conclusion. *In general, the quantum potential $U(t, x)$ is a generalized function (distribution). Therefore the price (as well as price change) trajectory is a generalized function (distribution) of the time variable t. Moreover, since the dynamical equation is nonlinear, one cannot guarantee even the existence of a solution.*

11.4 Classical and Quantum Financial Randomness

By considering singular quantum potentials we can model the Bohmian price dynamics *with trajectories having nonzero quadratic variation.* The main problem is that there are no existence theorems for such forces. Derivation of such theorems is an interesting mathematical problem, but it is completely outside of my expertise.

Another way to obtain a more realistic QL model for the financial market is to consider additional stochastic terms in Newton's equation for the price dynamics.

11.4.1 Randomness of Initial Conditions

Let us consider the financial Newton equation (11.1) with random initial conditions:

$$\frac{md^2q(t, \omega)}{dt^2} = f(t, q(t, \omega)) + g(t, q(t, \omega)), \tag{11.9}$$

$$q(0, \omega) = q_0(\omega), \quad \dot{q}(0, \omega) = \dot{q}_0(\omega), \tag{11.10}$$

where $q_0(\omega)$ and $\dot{q}_0(\omega)$ are two random variables giving the initial distribution of prices and price changes, respectively. This is the Cauchy problem for an ordinary differential equation depending on a parameter ω. If f satisfies the conditions of Theorem 11.1, i.e., both classical and quantum (behavioral) financial forces $f(t, q)$ and $g(t, q)$ are continuous and satisfy the Lipschitz condition with respect to the price variable q, then, for any ω, there exists the solution $q(t, \omega)$ having the class C^2 with respect to the time variable t. But through initial conditions the price depends on the random parameter ω, so $q(t, \omega)$ is a stochastic process. In the same way the price change $v(t, \omega) = \dot{q}(t, \omega)$ is also a stochastic process. These processes can be extremely complicated (through nonlinearity of coefficients f and g). In general, these are *nonstationary processes*. For example, the mathematical expectation $< q(t) >= Eq(t, \omega)$ and dispersion ("volatility") $\sigma^2(q(t)) = Eq^2(t, \omega) - < q(t) >^2$ can depend on t.

If at least one of the financial forces $f(t, x)$ and $g(t, x)$ is not continuous, then we consider the corresponding integral equations:

$$p(t, \omega) = p_0(\omega) + \int_{t_0}^{t} f(s, q(s, \omega))ds + \int_{t_0}^{t} g(s, q(s, \omega))ds, \tag{11.11}$$

$$q(t, \omega) = q_0(\omega) + \frac{1}{m}\int_{t_0}^{t} p(s, \omega)ds \tag{11.12}$$

Under the assumptions of Theorem 11.3 or Theorem 11.4, there exists a unique stochastic process with continuous trajectories, $q(t, \omega)$, $p(t, \omega)$, giving the solution of the system of integral equations (11.11), (11.12) with random initial conditions.

However, the trajectories still have zero quadratic variation. Therefore this model is not satisfactory.

11.4.2 Random Financial Mass

The parameter m, "financial mass", was considered as a constant of the model. In the real financial market m depends on t: $m \equiv m(t) = (m_1(t), \ldots, m_n(t))$. Here $m_j(t)$ is the volume of emission (the number of items) of shares of the jth corporation. Therefore the corresponding *market capitalization* is given by $T_j(t) = m_j(t)q_j(t)$.

In this way we modify the financial Newton equation (11.9): $m_j(t)\ddot{q}_j = f_j(t, q) + g_j(t, q)$. We set

$$F_j(t, q) = \frac{f_j(t, q) + g_j(t, q)}{m_j(t)}.$$

If these functions are continuous (e.g., $m_j(t) \geq \varepsilon_j > 0$ and continuous)[1] and satisfy the Lipschitz condition, then by Theorem 11.1 there exists a unique C^2-solution. If components $F_j(t, q)$ are discontinuous, but they satisfy the conditions of Theorem 11.3 or 11.4, then there exists a unique continuous solution of the corresponding integral equation with time-dependent financial masses. By considering the Bohmian model of the financial market with random initial conditions it is natural to assume that even the financial masses $m_j(t)$ are random variables, $m_j(t, \omega)$.

Thus the level of emission of jth share m_j depends on the classical state ω of the financial market: $m_j \equiv m_j(t, \omega)$. In this way we obtain the simplest stochastic modification of Bohmian dynamics:

$$\ddot{q}_j(t, \omega) = \frac{f_j(t, q(t, \omega)) + g_j(t, q(t, \omega))}{m_j(t, \omega)}$$

or in the integral version:

$$q_j(t, \omega) = q_{0j}(\omega) + \int_{t_0}^{t} v(s, \omega)ds, \tag{11.13}$$

$$v_j(t, \omega) = v_{0j}(\omega) + \int_{t_0}^{t} [f_j(s, q(s, \omega)) + g_j(s, q(s, \omega))]/m_j(s, \omega)ds. \tag{11.14}$$

If the financial mass can become zero at some moments of time, then *the price can have nonzero quadratic variation*. However, under such conditions *we do not have an existence theorem*.

11.5 Bohm–Vigier Stochastic Mechanics

The quadratic variation objection motivates consideration of the Bohm–Vigier stochastic model, instead of the completely deterministic Bohmian model. We follow here [40]. We recall that in the original Bohmian model the velocity of an individual particle is given by

$$v = \frac{\nabla S(q)}{m}. \tag{11.15}$$

---

[1] The condition $m_j(t) \geq \varepsilon_j > 0$ is very natural. To be accounted at the financial market, the volume of emission of any share should not be negligibly small.

If $\psi = Re^{iS/h}$, then Schrödinger's equation implies that

$$\frac{dv}{dt} = -\nabla(V + U),$$ (11.16)

where V and U are classical and quantum potentials respectively. In principle one can work only with the basic equation (11.15).

The basic assumption of Bohm and Vigier was that the velocity of an individual particle is given by

$$v = \frac{\nabla S(q)}{m} + \eta(t),$$ (11.17)

where $\eta(t)$ represents a random contribution to the velocity of that particle which fluctuates in a way that may be represented as a random process but with zero average. In the Bohm–Vigier model the stochastic mechanics quantum potential comes in through the average velocity and not the actual one.

We shall now apply the Bohm–Vigier model to the financial market, see also Haven [136]. Equation (11.17) is considered as the basic equation for the price velocity. Thus the real price becomes a random process (as well as in classical financial mathematics [281]). We can write the stochastic differential equation, SDE, for the price:

$$dq(t) = \frac{\nabla S(q)}{m} dt + \eta(t)dt.$$ (11.18)

To give rigorous mathematical meaning to the stochastic differential we assume that

$$\eta(t) = \frac{d\xi(t)}{dt},$$ (11.19)

for some stochastic process $\xi(t)$. Thus formally:

$$\eta(t)dt = \frac{d\xi(t)}{dt} dt = d\xi(t),$$ (11.20)

and the rigorous mathematical form of (11.18) is

$$dq(t) = \frac{\nabla S(q)}{m} dt + d\xi(t).$$ (11.21)

The expression (11.19) one can consider either formally or in the sense of distribution theory (we recall that for basic stochastic processes, e.g., the Wiener process, trajectories are not differentiable in the ordinary sense almost everywhere).

Suppose, for example, that the random contribution to the price dynamics is given by *white noise*, $\eta_{white\ noise}(t)$. It can be defined as the derivative (in the sense of distribution theory) of the Wiener process: $\eta_{white\ noise}(t) = dw(t)/dt$, thus

$$v = \frac{\nabla S(q)}{m} + \eta_{white\ noise}(t). \qquad (11.22)$$

In this case the price dynamics is given by the SDE

$$dq(t) = \frac{\nabla S(q)}{m} dt + dw(t). \qquad (11.23)$$

What is the main difference from the classical SDE description of the financial market? This is the presence of the pilot wave $\psi(t, q)$, the mental field of the financial market, which determines the coefficient of drift $\nabla S(q)/m$. Here $S \equiv S_\psi$. And the ψ-function is driven by a special field equation – Schrödinger's equation. The latter equation is not determined by the SDE (11.23). Thus, instead of one SDE, in the quantum-like model, we have the following system of two equations:

$$dq(t) = \frac{\nabla S_\psi(q)}{m} dt + d\xi(t). \qquad (11.24)$$

$$i\,h\frac{\partial \psi}{\partial t}(t, q) = -\frac{h^2}{2m}\frac{\partial^2 \psi}{\partial q^2}(t, q) + V(q)\psi(t, q). \qquad (11.25)$$

Finally we come back to the problem of the quadratic variation of the price. In the Bohm–Vigier stochastic model (for, e.g., white noise fluctuations of the price velocity) quadratic variation is nonzero.

11.6 Bohmian Model and Models with Stochastic Volatility

Some authors, see, e.g., [281] for details and references, consider the parameters of volatility $\sigma(t)$ as representing the market behavior. From such a point of view our financial wave $\psi(t, q)$ plays in the Bohmian financial model a role similar to the role of volatility $\sigma(t)$ in the standard stochastic financial models. We recall that the dynamics of $\psi(t, q)$ is driven by an independent equation, namely the Schrödinger equation, and $\psi(t, q)$ plays the role of a parameter of the dynamical equation for the price $q(t)$.

We recall the functioning of this scheme:

a) we find the financial wave $\psi(t, q)$ from the Schrödinger equation;
b) we find the corresponding quantum financial potential

$$U(t, q) \equiv U(t, q; \psi)$$

(it depends on ψ as a parameter);

c) we put $U(t, q; \psi)$ into the financial Newton equation through the quantum (behavioral) force $g(t, q; \psi) = -\partial U(t, q; \psi)/\partial q$.

We remark that conventional models with stochastic volatility work in the same way, see [281]. Here the price q_t is a solution of the stochastic differential equation

$$dq_t = q_t \mu(t, q_t, \sigma_t)dt + \sigma_t dw_t^\varepsilon, \tag{11.26}$$

where w_t^ε is the Wiener process and σ_t is a coefficient depending on time, price and volatility. And (this is a crucial point) volatility satisfies the following stochastic differential equation:

$$d\Delta_t = \alpha(t, \Delta_t)dt + b(t, \Delta_t)dw_t^\delta, \tag{11.27}$$

where $\Delta_t = \ln \sigma_t^2$ and w_t^δ is a Wiener process that is independent of w_t^ε.

One should first solve the equation for the volatility (11.27), then put σ_t into (11.26) and, finally, find the price q_t.

11.7 Classical and Quantum Contributions to Financial Randomness

As in conventional stochastic financial mathematics, see, e.g., [237, 281], we can interpret ω as representing a state of the financial market. The only difference is that in our model such an ω should be related to "classical state" of the financial market. Thus we interpret conventional randomness of the financial market as "classical randomness", i.e., randomness that is not determined by expectations of traders and other behavioral factors. Besides this "classical states" ω our model contains also "quantum states" ψ of the financial market describing the market's psychology. In fact all processes under consideration depend not only the classical state ω, but also on the quantum state ψ :

$$dv_j(t, \omega, \psi) = \frac{f_j(t, q(t, \omega, \psi), v(t, \omega, \psi), \omega)}{m_j(t, \omega)}dt + \frac{g_j(t, q(t, \omega, \psi), \omega, \psi)}{m_j(t, \omega)}dt$$
$$+ \sigma_j(t, \omega)dW_j(t, \omega). \tag{11.28}$$

We remark that the quantum force depends on the ψ-parameter even directly: $g_j = g_j(t, q, \omega, \psi)$. The initial condition for the SDE (11.28) depends only on ω : $q_j(0, \omega) = q_{j0}(\omega)$, $v_j(0, \omega) = v_{j0}(\omega)$. But in general the quantum state of the financial market is given not by the pure state ψ, but by the von Neumann density operator ρ. Therefore ψ in (11.28) is a quantum random parameter with the initial quantum probability distribution given by the density operator at the initial moment: $\rho(0) = \rho_0$. We recall that the Schrödinger equation for the pure state implies the von Neumann equation for the density operator:

$$i\dot\rho(t) = [\hat H, \rho]. \tag{11.29}$$

Chapter 12
Appendix

12.1 Independence

12.1.1 Kolmogorovian Model

This section is a complement to Section 2.1.2 of Chapter 2.

By Kolmogorov's axiomatics two events $A, B \in \mathcal{F}$ are *independent* if

$$\mathbf{P}(A \cap B) = \mathbf{P}(A)\mathbf{P}(B). \tag{12.1}$$

Consequently, two random variables a and b are independent if

$$p^{ab}(\alpha, \beta) = p^a(\alpha)p^b(\beta). \tag{12.2}$$

If two random variables are independent, then

$$Eab = EaEb \tag{12.3}$$

and by (2.8)

$$\text{cov}(a, b) = 0. \tag{12.4}$$

Hence, if two random variables are independent, then their covariation is equal to zero. However, in general the latter does not imply the former!

We remark that Kolmogorov's definition of independence involves the joint probability distribution of two random variables. It is not always possible in the coming non-Kolmogorovian considerations to define it – in QM and my contextual probabilistic model, Chapter 3. Let us write the condition of independence (12.2) by using only transition probabilities. The latter can be defined even in non-Kolmogorovian models.

By using Bayes formula and (12.2) we obtain for two independent nondegenerate (see Section 2.1.2) random variables

$$p_{\beta|\alpha} \equiv \mathbf{P}(b = \beta | a = \alpha) = p^b(\beta) \tag{12.5}$$

for all values $a = \alpha$. Consider two successive measurements, first of a and then of b; then the result of the latter is the same as if a was not done. In the same way

$$p_{\alpha|\beta} \equiv \mathbf{P}(a = \alpha | b = \beta) = p^a(\alpha). \tag{12.6}$$

Now let a and b satisfy conditions (12.5) or (12.6). Then (12.2) holds.

Remark 12.1 Conditions (12.5) and (12.6) are equivalent. This fact reflects that in the Kolmogorovian model the relation of independence is *symmetric*. If b is independent of a – this fact is expressed by (12.5) – then a is independent of b – as expressed by (12.6) – and vice versa. In principle, in real life it is possible to imagine the situation of *nonsymmetric dependence*. For example, b does not depend on a, i.e., (12.5) holds, but a depends on b, i.e., (12.6) does not hold. Such a situation is not described by the Kolmogorovian model, see Chapter 3.

Remark 12.2 The real "physical" meaning of independence is encoded in the pair of equalities (12.5) and (12.6). Equality (12.2) is just a nice mathematical definition unifying (for nondegenerate random variables) the pair of equalities (12.5), (12.6).

Conditions (12.5) and (12.6) imply that

$$p_{\beta|\alpha_1} = \ldots p_{\beta|\alpha_n} = \ldots = c_\beta \text{ for any } \beta \in X_b; \tag{12.7}$$

$$p_{\alpha|\beta_1} = \ldots p_{\alpha|\beta_n} = \ldots = c_\alpha \text{ for any } \alpha \in X_a, \tag{12.8}$$

where c_β and c_α are nonnegative constants.

On the other hand, either of conditions (12.7) and (12.8) implies (12.2) and, hence, both (12.5) and (12.6). For example, let condition (12.7) hold. Thus $p^{ab}(\alpha, \beta) = p^a(\alpha)p_{\beta|\alpha} = p^a(\alpha)c_\beta$. But we have $p^b(\beta) = \sum_\alpha p^{ab}(\alpha, \beta) = c_\beta \sum_\alpha p^a(\alpha) = c_\beta$. So, in the Kolmogorovian model

$$c_\alpha = p^a(\alpha), \quad c_\beta = p^b(\beta). \tag{12.9}$$

We remark that formally conditions (12.7) and (12.8) are weaker than conditions (12.5) and (12.6). For example, "physically" (12.7) means that if first measurement of a has been performed and then measurement of b is performed, then the *result* of the b-measurement does not depend on the *result* of the preceding a-measurement.[1] However, as we have seen, in the Kolmogorovian model all these conditions are equivalent.

12.1.2 Quantum Model

This section is a complement to Section 2.4.

---

[1] A priori $c_\beta = p_{\beta|\alpha_1} = \ldots p_{\beta|\alpha_n} = \ldots$ might be different from $p^b(\beta)$; this c_β might depend on a (in general), but not on its concrete values.

We would like to generalize the notion of independence to quantum observables with nondegenerate spectra.[2] Two noncommuting quantum observables are considered as complementary and they cannot be jointly measured – at least by the conventional interpretation of QM. Therefore Kolmogorov's original definition (12.2) cannot be directly generalized to QM. However, its reformulation in terms of transition probabilities, (12.7) and (12.8), can be easily transformed to QM.

We call two quantum observables *independent* if, e.g., (12.7) holds. It implies that $c_\beta = p_{\beta|\alpha_1} = p_{\beta|\alpha_2} = \ldots$. Since in QM the matrix $\mathbf{P}^{b|a}$ is doubly stochastic, we have $\sum_\alpha p_{\beta|\alpha} = 1$. This implies, first of all, that in the infinite-dimensional case there are no independent observables (with discrete spectrum!). In the finite-dimensional case independence is equivalent to the condition

$$|\langle e_\beta^b, e_\alpha^a \rangle|^2 = \frac{1}{n}, \tag{12.10}$$

where n is the dimension of Hilbert space. Two orthonormal bases that satisfy (12.10) are called *mutually unbiased bases*. To describe such bases is an extremely complex geometric problem, see e.g., Bengtsson [33] for details and literature.

We remark that the relation of independence in QM is symmetric, in the same way as in classical probability theory. Condition (12.7) is equivalent to condition of (12.10) and the latter is also equivalent to condition (12.8). But, of course, condition (12.9) is violated (for some state ψ).

We remark that one can properly define covariance of two (even noncommuting) quantum observables (in spite of complementarity). However, we will not do this specially for quantum probabilities. See Chapter 3 for a definition of covariance in the general contextual probabilistic model and Section 12.4 of this Appendix for embedding of the quantum model in it.

12.1.3 Växjö Model

This section is a complement to Section 3.1.4.

As in QM, to define independent observables, we use conditions (12.7) and (12.8) However, the relation of independence is not symmetric (unlike the Kolmogorovian and quantum models). In general, conditions (12.7) and (12.8) are not equivalent. We say that *b is independent of a* if (12.7) holds; *a is independent of b* if (12.8) holds.

We remark that, for coefficients c_β defined by (12.7), $\sum_\beta c_\beta = \sum_\beta p_{\beta|\alpha_1} = 1$, since the matrix $\mathbf{P}^{b|a}$ is always stochastic; in the same way, for coefficients c_α defined by (12.8), their sum is equal to 1.

---

[2] An operator with nondegenerate spectrum is an analogue of a nondegenerate random variable, see (2.14).

Proposition 12.1 *Let b not depend on a. Then* $\mathrm{cov}_C(b|a) = 0$.

Proof We have, see (3.9) in Section 3.1.4:

$$\mathrm{cov}_C(b|a) = \sum_{\alpha,\beta} \alpha\beta p_C^a(\alpha) p_{\beta|\alpha} + \bar{a}_C\bar{b}_C - \bar{b}_C \sum_{\alpha,\beta} \alpha p_C^a(\alpha) p_{\beta|\alpha} - \bar{a}_C \sum_{\alpha,\beta} \beta p_C^a(\alpha) p_{\beta|\alpha}$$

$$= \sum_{\alpha,\beta} \alpha\beta p_C^a(\alpha) c_\beta + \bar{a}_C\bar{b}_C - \bar{b}_C \sum_{\alpha,\beta} \alpha p_C^a(\alpha) c_\beta - \bar{a}_C \sum_{\alpha,\beta} \beta p_C^a(\alpha) c_\beta$$

$$= \bar{a}_C \sum_\alpha \beta c_\beta + \bar{a}_C\bar{b}_C - \bar{a}_C\bar{b}_C \sum_\beta c_\beta - \bar{a}_C \sum_\alpha p_C^a(\alpha) \sum_\beta \beta c_\beta = 0.$$

Consider two observables a and b taking n values. Let $\mathbf{P}^{b|a}$ be doubly stochastic and let b be independent of a, i.e., (12.7) holds. Then, as in QM, we get that, for each β, $1 = \sum_\alpha p_{\beta|\alpha} = nc_\beta$. Thus, $c_\beta = 1/n$. In the same way if a is independent of b and $\mathbf{P}^{a|b}$ is doubly stochastic, then all coefficients $c_\alpha = 1/n$.

Now let a and b be symmetrically conditioned, **CS**, see (2.18) in Chapter 2. Then both matrices $\mathbf{P}^{b|a}$ and $\mathbf{P}^{a|b}$ are doubly stochastic. Let b be independent of a. Then all coefficients $c_\beta = 1/n$. Hence, $p_{\alpha|\beta} = p_{\beta|\alpha} = 1/n$. Thus a is independent of b.

We remark that under condition **SC** one can construct a representation of observables by self-adjoint operators in the complex Hilbert space. This representation has all features of the conventional quantum representation; see the discussion in the introduction to Chapter 4.

12.2 Proof of Wigner's Inequality

This subsection is a complement to Section 2.2.2.

Let $\mathcal{P} = (\Lambda, \mathcal{F}, \mathbf{P})$ be a probability space. We remark that in physics people (following Bell) use typically the symbol Λ, instead of Ω. It is interpreted as a set of hidden parameters: $\lambda \in \Lambda$ determine the "prequantum state" of a quantum system. Such a state is not described by quantum formalism. If one were able to approach this state, it would be possible to determine the values of all quantum observables.

Proof (Wigner's inequality) We have:

$$\mathbf{P}(a_1(\lambda) = +1, a_2(\lambda) = +1)$$

$$= \mathbf{P}(a_1(\lambda) = +1, a_2(\lambda) = +1, a_3(\lambda) = +1) \qquad (12.11)$$

$$+ \mathbf{P}(a_1(\lambda) = +1, a_2(\lambda) = +1, a_3(\lambda) = -1),$$

$$\mathbf{P}(a_2(\lambda) = -1, a_3(\lambda) = +1)$$

$$= \mathbf{P}(a_1(\lambda) = +1, a_2(\lambda) = -1, a_3(\lambda) = +1) \tag{12.12}$$

$$+ \mathbf{P}(\lambda \in \Lambda : a_1(\lambda) = -1, a_2(\lambda) = -1, a_3(\lambda) = +1),$$

and

$$\mathbf{P}(a_1(\lambda) = +1, a_3(\lambda) = +1)$$

$$= \mathbf{P}(a_1(\lambda) = +1, a_2(\lambda) = +1, a_3(\lambda) = +1) \tag{12.13}$$

$$+ \mathbf{P}(a_1(\lambda) = +1, a_2(\lambda) = -1, a_3(\lambda) = +1).$$

If we add together the equations (12.11) and (12.12) we obtain

$$\mathbf{P}(a_1(\lambda) = +1, a_2(\lambda) = +1) + \mathbf{P}(a_2(\lambda) = -1, a_3(\lambda) = +1)$$

$$= \mathbf{P}(a_1(\lambda) = +1, a_2(\lambda) = +1, a_3(\lambda) = +1)$$

$$+ \mathbf{P}(a_1(\lambda) = +1, a_2(\lambda) = +1, a_3(\lambda) = -1) \tag{12.14}$$

$$+ \mathbf{P}(a_1(\lambda) = +1, a_2(\lambda) = -1, a_3(\lambda) = +1)$$

$$+ \mathbf{P}(a_1(\lambda) = -1, a_2(\lambda) = -1, a_3(\lambda) = +1).$$

But the first and the third terms on the right hand side of this equation are just those which when added together make up the term $\mathbf{P}(a_1(\lambda) = +1, c(\lambda) = +1)$ (Kolmogorov probability is additive). It therefore follows that

$$\mathbf{P}(a_1(\lambda) = +1, a_2(\lambda) = +1) + \mathbf{P}(a_2(\lambda) = -1, c(\lambda) = +1)$$

$$= \mathbf{P}(a_1(\lambda) = +1, c(\lambda) = +1)$$

$$+ \mathbf{P}(a_1(\lambda) = +1, a_2(\lambda) = +1, c(\lambda) = -1) \tag{12.15}$$

$$+ \mathbf{P}(a_1(\lambda) = -1, a_2(\lambda) = -1, c(\lambda) = +1)$$

By using nonnegativity of probability we obtain the inequality

$$\mathbf{P}(a_1(\lambda) = +1, a_2(\lambda) = +1) + \mathbf{P}(a_2(\lambda) = -1, c(\lambda) = +1)$$

$$\geq \mathbf{P}(a_1(\lambda) = +1, c(\lambda) = +1) \tag{12.16}$$

12.3 Projection Postulate

This section is a complement to Sections 3.1.2 and 2.4.

We did not include the von Neumann's projection postulate [301] in the main list of postulates of quantum mechanics, because one can proceed quite far without this postulate.

Projection Postulate. *Let* a *be a physical observable represented by a self-adjoint operator* $\hat{a}$ *having a purely discrete nondegenerate spectrum. Any measurement of the observable* a *on the quantum state* ψ *induces transition from the state* ψ *into one of eigenvectors* e_k^a *of the operator* $\hat{a}$.

See von Neumann [301], p. 216: "Under the above assumption on $\hat{a}$, a measurement of a the has the consequence of changing each state ψ into one of the states $e_1^a, e_2^a, \ldots$ which are connected with respective results of measurement $\alpha_1, \alpha_2, \ldots$ The probabilities of these changes are therefore equal to the measurement probabilities for $\alpha_1, \alpha_2, \ldots$"

By the Copenhagen interpretation the ψ-function gives the state of an individual quantum system. Therefore von Neumann's postulate is about the result of transformation of the state of, e.g., an electron in the process of measurement. By the ensemble interpretation the ψ-function describes the state of an ensemble of particles. Here von Neumann's postulate describes the state of the post-measurement ensemble created by selection of particles with the fixed result of measurement $a = \alpha_k$.

The original von Neumann's postulate was formulated only for observables represented by operators with nondegenerate spectra. This fact is practically forgotten; von Neumann pointed out that if the spectrum is degenerate, measurement induces not a pure quantum state (fixed wave function $\psi_{post-meas}$), but a mixture of pure states. Later Lüders "generalized" von Neumann's projection postulate to operators with degenerate spectra. By Lüders the post-measurement state is always a pure state again. It is the projection of the original state ψ to the subspace of eigenvectors corresponding to the eigenvalue, say α_k, obtained as the result of measurement, i.e., $a = \alpha_k$. In modern literature on QM Lüders' postulate is considered as simply a generalization of von Neumann's postulate; people are not able to see the obvious contradiction between them, cf. [210, 212].

12.4 Contextual View of Kolmogorov and Quantum Models

12.4.1 Contextual Models Induced by the Classical (Kolmogorov) Model

This subsection is a complement to Sections 2.1 and 3.1.

We start with Kolmogorov's model. Let $\mathcal{P} = (\Omega, \mathcal{F}, \mathbf{P})$ be a Kolmogorov probability space. It induces various Växjö models through various choices of collections of contexts $\mathcal{C}$ and observables $\mathcal{O}$ as well as systems of sets representing selection contexts for values of observables from $\mathcal{O}$. The collection of contexts $\mathcal{C}$ can be chosen as some sub-family of $\mathcal{F}$ consisting of sets of positive probability: $\mathbf{P}(C) > 0$, $C \in \mathcal{C}$. The crucial point is that the collection of contexts need not form a σ-algebra or algebra. The collection of observables $\mathcal{O}$ can be chosen as a set of nondegenerate random variables.

For a discrete variable a, its essential range of values ("spectrum") is given by the set $X_a = \{\alpha\}$, where $\mathbf{P}(\omega \in \Omega : a(\omega) = \alpha) > 0$.

Contextual probabilities in Växjö models induced by the Kolmogorov model are given by the Bayes' formula (so, these are simply conditional probabilities). For an observable (random variable) a and its value $\alpha \in X_a$ the $[a = \alpha]$-selection context C_α is given by the set $C_\alpha^a = \{\omega \in \Omega : a(\omega) = \alpha\}$. The condition (3.5) evidently holds.

12.4.2 Contextual Models Induced by the Quantum (Dirac–von neumann) Model

This subsection is a complement to Sections 2.3 and 3.1.

We now consider Växjö models induced by the quantum model. The set of contexts $\mathcal{C}$ can be chosen as a subset of the unit sphere S of complex Hilbert space $\mathcal{H}$ (a collection of normalized vectors[3]): each context $C \in \mathcal{C}$ is encoded by a vector $\psi \in S : C \equiv C_\psi$. Each Växjö model induced by QM is based on its own subset of vectors $\psi \in S$. The maximal set of contexts is given by the whole S. It is used in QM.

The set of observables $\mathcal{O}$ can be chosen as a subset of the space of self-adjoint operators having purely discrete spectra.[4] Contextual probabilities are defined by Born's rule. We consider the simplest case: all operators belonging to $\mathcal{O}$ have nondegenerate spectra. In this case Born's rule has the form (2.34). Let an operator $\widehat{a} \in \mathcal{O}$ have the spectrum $X_a = \{\alpha_1, \ldots, \alpha_N, \ldots\}$, $\alpha_i \neq \alpha_j$ and let e_α^a, $\alpha \in X_a$, be the corresponding eigenvectors. Then $\mathbf{P}(a = \alpha_i | C_\psi) = |\langle \psi, e_{\alpha_i}^a \rangle|^2$. The $[a = \alpha]$-selection contexts C_α are represented by the eigenvectors $C_\alpha \equiv C_{e_\alpha^a}$.

We have $\mathbf{P}(b = \beta | C_\alpha) = \mathbf{P}(a = \alpha | C_\beta) = |\langle e_\alpha^a, e_\beta^b \rangle|^2$.

12.5 Generalization of Quantum Formalism

Let us consider a finite-dimensional Hilbert space $\mathcal{H}$. Let (for primary consideration) $\mathcal{E} = \{e_j\}_{j=1}^n$ be an *orthonormal basis*

---

[3] Another way to describe quantum probabilities within the contextual probabilistic model is to proceed (similarly to the Kolmogorov case) by representing contexts not by single vectors from the unit sphere S, but by equivalence classes of these vectors: ψ_1 is equivalent to ψ_2 if $\psi_1 = c\psi_2$, where $|c| = 1$.

[4] It is possible to generalize the contextual probabilistic model to cover observables with "continuous spectra." However, we do not need such a model for the coming applications.

$$\psi = \sum_j u_j e_j, u_j = u_j(\psi) \in \mathbf{C}. \tag{12.17}$$

Each $\mathcal{E}$ generates a class of (conventional) quantum observables, self-adjoint operators, see [301, 90]:

$$\hat{a}\psi = \sum_j \alpha_j u_j(\psi) e_j, \tag{12.18}$$

where $X_a = \{\alpha_1, \ldots, \alpha_n\}, \alpha_j \in \mathbf{R}, \alpha_j \neq \alpha_i$ is the range of values of a (so we start with consideration of observables with nondegenerate spectra).

Let now $\mathcal{E} = \{e_j\}_{j=1}^n$ be an arbitrary basis (thus in general $\langle e_j, e_i \rangle \neq 0, i \neq j$) consisting of normalized vectors, i.e., $\langle e_j, e_j \rangle = 1.$[5] Let us generalize the Dirac–von Neumann formalism by considering observables (12.18) for an arbitrary (in general nonorthogonal) basis $\mathcal{E}$. We consider an *arbitrary nonzero vector* of $\mathcal{H}$ as a pure quantum state. Thus the condition $\psi \in S$, where S is the unit sphere is eliminated from the model. We postulate – generalizing Born's postulate – that

$$\mathbf{P}_\psi(a = \alpha_j) = \frac{|u_j(\psi)|^2}{\sum_j |u_j(\psi)|^2}, \tag{12.19}$$

where the coefficients $u_j(\psi)$ are given by the expansion (12.17).

If $\mathcal{E}$ is an orthonormal basis, then $u_j(\psi) = \langle \psi, e_j \rangle, \sum_j |u_j(\psi)|^2 = \|\psi\|^2$ and for a normalized vector ψ, we obtain the ordinary Born's rule. Our generalization of the Dirac–von Neumann formalism is very close to another well–known (and very popular in quantum information) generalization of the class of quantum observables, namely, to the formalism of a *positive operator valued measure* (POVM), [46, 148].

To proceed further in this way, we introduce (in general nonorthogonal) projectors on the basis vectors: $\Pi_j \psi = u_j(\psi) e_j$. We remark that $\Pi_j^2 = \Pi_j$, but in general $\Pi_j^* \neq \Pi_j$. We have: $|u_j(\psi)|^2 = \langle \Pi_j \psi, \Pi_j \psi \rangle = \langle M_j \psi, \psi \rangle$, where $M_j = \Pi_j^* \Pi_j$. We remark that each M_j is self-adjoint and, moreover, positively defined. We also set $M = \sum_j M_j$. Then our generalization of Born's rule can be written as

$$\mathbf{P}_\psi(a = \alpha_j) = \frac{\langle M_j \psi, \psi \rangle}{\langle M \psi, \psi \rangle} = \frac{\mathrm{Tr}\, \rho_\psi M_j}{\mathrm{Tr}\, \rho_\psi M}, \tag{12.20}$$

where $\rho_\psi = \psi \otimes \psi$. We remark that, for an arbitrary nonzero ψ, the operator $\rho_\psi \geq 0$.

Now we generalize the conventional notion of the density operator, by considering any nonzero $\rho \geq 0$ as a generalized density operator (we recall that at the

---

[5] We remark that QLRA, Section 4.2, produces the a-basis with normalized vectors, $\|e_\alpha^a\|^2 = 1$. This is a consequence of stochasticity of an arbitrary matrix of "transition probabilities" (which was used by QLRA to produce the a-basis). Thus we consider now a purely linear algebraic version of this situation.

moment we consider a finite-dimensional space). The corresponding generalization
of Born's postulate has the following form:

$$\mathbf{P}_\psi(a = \alpha_j) = \frac{\mathrm{Tr}\, \rho\, M_j}{\mathrm{Tr}\, \rho\, M}.$$ (12.21)

The only difference from the POVM formalism is that the operator $M \neq I$ (the unit
operator).

We remark that $\langle M\psi, \psi \rangle = \sum_j |u_j(\psi)|^2 \neq 0, \psi \neq 0$. Thus (we are in the
finite-dimensional case) the inverse operator M^{-1} is well defined.

We now proceed with our formalization and consider an arbitrary (separable)
Hilbert space $\mathcal{H}$.

Definition 12.1 *A generalized quantum state is represented by an arbitrary trace
class nonnegative (nonzero) operator* $\rho : \rho \geq 0, 0 < \mathrm{Tr}\rho < \infty$.

Definition 12.2 *A generalized quantum observable is represented by an arbitrary
(so in general nonnormalized) positive operator valued measure E on a measurable
space $(X, \mathcal{F})$ such that $E(X) > 0$.*

Thus, for a generalized quantum observable E, we have

1) $E(B) \geq 0$, for any set $B \in \mathcal{F}$, and $E(X) > 0$;
2) $E(\cup_{j=1}^n B_j) = \sum_{j=1}^n E(B_j)$ for all disjoint sequences $\{B_j\}$ in $\mathcal{F}$.

Generalized Born's rule: Let ρ and E be a generalized quantum state and
observable, respectively. Then the probability of finding the result x of the
E-measurement in a measurable set B (for an ensemble represented by ρ) is given
by

$$\mathbf{P}_\rho(x \in B) = \frac{\mathrm{Tr}\rho\, E(B)}{\mathrm{Tr}\rho\, E(X)}.$$ (12.22)

We remark that $\mathrm{Tr}\rho\, E(X) > 0$. To prove this, we consider the spectral expansion
of the trace class operator $\rho = \sum_j q_j \psi_j \otimes \psi_j$. Here at least one $q_j > 0$. Then
$\mathrm{Tr}\rho\, E(X) = \sum_j q_j \langle E(X)\psi_j, \psi_j \rangle > 0$.

We now come back to the model considered at the beginning of this section:
a finite-dimensional space. We would like to model in the abstract linear algebra
framework the situation considered in Section 4.2, but in the case of the (in gen-
eral) non-doubly stochastic matrix of $b|a$-contextual probabilities $\mathbf{P}^{b|a}$, i.e., for a
nonorthogonal basis $\{e_\alpha^a\}$. We consider two observables, one is a conventional self-
adjoint operator $\hat{b}$ and the other is a generalized observable $\hat{a}$. Thus the b-basis
$\mathcal{E}^b = \{e_j^b\}$ is orthonormal, but the a-basis $\mathcal{E}^a = \{e_j^a\}$ need not be (but we emphasize
that even the latter is normalized). Any vector e_j^a is a conventional (pure) quan-
tum state. Thus by the rules of conventional QM we can find "transition proba-
bilities": $p^{b|a}(\beta_i|\alpha_j) = \mathbf{P}_{e_j^a}(b = \beta_i) = |\langle e_j^a, e_i^b \rangle|^2$. Since $\mathcal{E}^b$ is orthonormal, we
have: $\sum_i p^{b|a}(\beta_i|\alpha_j) = \sum_i |\langle e_j^a, e_i^b \rangle|^2 = \|e_j^a\|^2 = 1$. The matrix of $b|a$-transition

probabilities $P^{b|a}$ is stochastic (as it should be). However, if $\mathcal{E}^a$ is not orthonormal, then $P^{b|a}$ is not doubly stochastic.

On the other hand, we can expand each e_i^b with respect to $\mathcal{E}^a$: $e_i^b = \sum_j u_j(e_i^b)e_j^a$. By our generalized Born's rule: $p^{a|b}(\alpha_j|\beta_i) = \mathbf{P}_{e_i^b}(a = \alpha_j) = |u_j(e_i^b)|^2/\sum_j |u_j(e_i^b)|^2$. We have $\sum_j p^{a|b}(\alpha_j|\beta_i) = 1$. Thus even the matrix of transition probabilities $P^{a|b}$ is stochastic.

Finally, we remark that all previous considerations are valid even in the case when both observables are generalized.

12.6 Bohmian Mechanics

In Bohmian mechanics (which is a completion of quantum mechanics) a quantum particle has well-defined position and momentum. Thus it has a trajectory (which is totally impossible by the Copenhagen interpretation). The crucial point is that a particle is guided by a so-called pilot wave, which is given by the wave function. Since the wave function of a family of N particles satisfies Schrödinger's equation not on "physical space" $\mathbf{R}^3$, but on the configuration space $\mathbf{R}^{3N}$, and since the equation of motion of any particle in general depends nontrivially on the wave function of the whole family, the Bohmian model is nonlocal.[6] Precisely this problem was the main reason for Schrödinger to reject his original interpretation of the wave function as a physical wave and to switch to Born's probabilistic interpretation.

We now present the detailed derivation of the equations of motion of a quantum particle in the Bohmian model of quantum mechanics. The dynamics of the wave function $\psi(t, q)$ is described by Schrödinger's equation (so this equation was simply borrowed from the conventional quantum formalism):

$$i\,\hbar\frac{\partial\psi}{\partial t}(t, q) = -\frac{\hbar^2}{2m}\frac{\partial^2\psi}{\partial q^2}(t, q) + V(t, q)\psi(t, q). \tag{12.23}$$

Here $\psi(t, q)$ is a complex-valued function. At the moment we prefer not to discuss the conventional probabilistic interpretation of $\psi(t, q)$. We consider $\psi(t, q)$ as just a field.[7] We consider the one-dimensional case, but the generalization to the multidimensional case, $q = (q_1, \ldots, q_n)$, is straightforward. Let us write the wave function $\psi(t, q)$ in the following form:

$$\psi(t, q) = R(t, q)e^{i\frac{S(t,q)}{\hbar}}, \tag{12.24}$$

---

[6] It seems that the latter feature was unacceptable for Albert Einstein, who considered Bohmian mechanics as a "cheap solution" of the problem of completion of quantum mechanics.

[7] We recall that by the probability interpretation of $\psi(t, q)$ (which was proposed by Max Born) the quantity $|\psi(t, q)|^2$ gives the probability of finding a quantum particle at the point q at the moment t, see (2.38), Section 2.3.1.

where $R(t, q) = |\psi(t, q)|$ and $\theta(t, q) = S(t, q)/\hbar$ is the argument of the complex number $\psi(t, q)$.

We put (12.24) into Schrödinger's equation (12.23). We have

$$i\hbar\frac{\partial\psi}{\partial t} = i\hbar\left(\frac{\partial R}{\partial t}e^{iS/\hbar} + \frac{iR}{\hbar}\frac{\partial S}{\partial t}e^{iS/\hbar}\right) = i\hbar\frac{\partial R}{\partial t}e^{iS/\hbar} - R\frac{\partial S}{\partial t}e^{iS/\hbar}$$

and

$$\frac{\partial\psi}{\partial q} = \frac{\partial R}{\partial q}e^{iS/\hbar} + \frac{iR}{\hbar}\frac{\partial S}{\partial q}e^{iS/\hbar}$$

and hence

$$\frac{\partial^2\psi}{\partial q^2} = \frac{\partial^2 R}{\partial q^2}e^{iS/\hbar} + \frac{2i}{\hbar}\frac{\partial R}{\partial q}\frac{\partial S}{\partial q}e^{iS/\hbar} + \frac{iR}{\hbar}\frac{\partial^2 S}{\partial q^2}e^{iS/\hbar} - \frac{R}{\hbar^2}\left(\frac{\partial S}{\partial q}\right)^2 e^{iS/\hbar}$$

We obtain the differential equations

$$\frac{\partial R}{\partial t} = \frac{-1}{2m}\left(2\frac{\partial R}{\partial q}\frac{\partial S}{\partial q} + R\frac{\partial^2 S}{\partial q^2}\right),\tag{12.25}$$

$$-R\frac{\partial S}{\partial t} = -\frac{\hbar^2}{2m}\left(\frac{\partial^2 R}{\partial q^2} - \frac{R}{\hbar^2}\left(\frac{\partial S}{\partial q}\right)^2\right) + VR.\tag{12.26}$$

By multiplying the right and left-hand sides of the equation (12.25) by $2R$ and using the trivial equalities

$$\frac{\partial R^2}{\partial t} = 2R\frac{\partial R}{\partial t}$$

and

$$\frac{\partial}{\partial q}(R^2\frac{\partial S}{\partial q}) = 2R\frac{\partial R}{\partial q}\frac{\partial S}{\partial q} + R^2\frac{\partial^2 S}{\partial q^2},$$

we derive the equation for R^2:

$$\frac{\partial R^2}{\partial t} + \frac{1}{m}\frac{\partial}{\partial q}\left(R^2\frac{\partial S}{\partial q}\right) = 0.\tag{12.27}$$

We remark that if one uses Born's probabilistic interpretation of the wave function, then

$$R^2(t, x) = |\psi(t, x)|^2$$

gives the probability. Thus (12.27) is the equation describing the dynamics of the probability distribution (in physics it is called the continuity equation).

The second equation can be written in the form

$$\frac{\partial S}{\partial t} + \frac{1}{2m}\left(\frac{\partial S}{\partial q}\right)^2 + \left(V - \frac{\hbar^2}{2mR}\frac{\partial^2 R}{\partial q^2}\right) = 0. \tag{12.28}$$

Suppose that

$$\frac{\hbar^2}{2m} << 1$$

and that the contribution of the term

$$\frac{\hbar^2}{2mR}\frac{\partial^2 R}{\partial q^2}$$

can be neglected. Then we obtain the equation

$$\frac{\partial S}{\partial t} + \frac{1}{2m}\left(\frac{\partial S}{\partial q}\right)^2 + V = 0. \tag{12.29}$$

From classical mechanics, we know that this is the classical Hamilton–Jacobi equation, which corresponds to the dynamics of particles:

$$p = \frac{\partial S}{\partial q} \quad \text{or} \quad m\dot{q} = \frac{\partial S}{\partial q}, \tag{12.30}$$

where particles move normal to the surface $S = const$.

David Bohm proposed that the equation (12.28) that should be interpreted in the same way. But we see that in this equation the classical potential V is perturbed by an additional "quantum potential"

$$U = \frac{\hbar^2}{2mR}\frac{\partial^2 R}{\partial q^2}.$$

Thus in Bohmian mechanics the motion of a particle is described by the usual Newton equation, but with the force corresponding to the combination of the classical potential V and the quantum one U:

$$m\frac{dv}{dt} = -\left(\frac{\partial V}{\partial q} - \frac{\partial U}{\partial q}\right). \tag{12.31}$$

The crucial point is that the potential U is itself driven by a field equation – Schrödinger's equation (12.23). Thus the equation (12.31) cannot be considered as just the Newtonian classical dynamics (because the potential U depends on ψ as a field parameter). We shall call (12.31) the *Bohm–Newton equation*.

We remark that typically in books on Bohmian mechanics [40, 149] it is emphasized that equation (12.31) is nothing other than the ordinary Newton equation. This creates the impression that the Bohmian approach gives the possibility of reducing quantum mechanics to ordinary classical mechanics. However, this is not the case. Equation (12.31) does not provide the complete description of the dynamics of a system. Since, as was pointed out, the quantum potential U is determined through the wave function ψ and the latter evolves according to the Schrödinger equation, the dynamics given by the Bohm–Newton equation cannot be considered independently of Schrödinger's dynamics.

References

1. Accardi, L.: Can mathematics help solving the interpretation problem of quantum theory? Nuovo Cimento B 110, 685–721 (1995)
2. Accardi, L.: Foundations of Quantum Mechanics: a quantum probabilistic approach, In: Tarozzi, G. and van der Merwe, A. (eds.) *The Nature of Quantum Paradoxes*, (Springer, Berlin), pp. 257–323 (1988). Preprint Dipartimento di Matematica, Università di Roma Tor Vergata (1986)
3. Accardi, L.: Could we now convince Einstein? In: Adenier, G., Khrennikov, A. Yu. and Nieuwenhuizen, T. M. (eds.) *Quantum Theory: Reconsideration of Foundations - 3*, AIP Conf. Proc., Vol. 810 (American Institute of Physics, Melville, NY 2006), pp. 3–18
4. Accardi, L. and Khrennikov, A. Yu.: Chameleon effect, the range of values hypothesis and reproducing the EPR-Bohm correlations. In: Adenier, G., Fuchs, C. and Khrennikov, A. Yu. (eds.) *Foundations of Probability and Physics - 4*. AIP Conf. Proc., Vol. 889 (American Institute of Physics, Melville, NY 2007), pp. 21–29
5. Adenier, G., Khrennikov, A. Yu. and Nieuwenhuizen, T. M. (eds.): *Quantum Theory: Reconsideration of Foundations - 3*. AIP Conf. Proc., Vol. 810 (American Institute of Physics, Melville, NY 2006)
6. Adenier, G., Fuchs, C. and Khrennikov, A. Yu. (eds.): *Foundations of Probability and Physics - 3*. AIP Conf. Proc., Vol. 889 (American Institute of Physics, Melville, NY 2007)
7. Adenier, G. and Khrennikov, A. Yu.: Is the fair sampling assumption supported by EPR experiments? J. Phys. B: At., Mol. Opt. Phys. 40, 131–141 (2007)
8. Aerts, D. and Aerts, S.: Applications of quantum statistics in psychological studies of decision processes. Found. Sci. 1, 85–97 (1995)
9. Aerts, D., D'Hooghe, B., Posiewnik, A., Pykacz, J., Dehaene, J. and De Moor, B.: How to play two-player restricted quantum games with 10 cards. Int. J. Theor. Phys. 47, 61–68 (2008)
10. Albert, D. Z. and Loewer, B.: Interpreting the many worlds interpretation. Synthese 77, 195–213 (1988)
11. Albert, D. Z.: *Quantum Mechanics and Experience* (Harvard University Press, Cambridge, MA 1992)
12. Albeverio, S., Khrennikov, A. Yu. and Kloeden, P. E.: Memory retrieval as a p-adic dynamical system. Biosystems 49, 105–115 (1999)
13. Allahverdyan, A., Khrennikov, A. Yu. and Nieuwenhuizen, T. M.: Brownian entanglement. Phys. Rev. A 71, 032102-1 – 14 (2005)
14. Amit, D. J.: *Modeling Brain Function* (Cambridge University Press, Cambridge 1989)
15. Amit, D. J.: Attractor neural networks and biological reality: associative memory and learning. Future Generation Comput. Syst. 6, 111–119 (1990)
16. Anderson, J. R.: *How Can the Human Mind Occur in the Physical Universe?* (Oxford University Press, New York 2007)

17. Andreev, V. A. and Man'ko, V. I.: Quantum tomography and verification of generalized Bell-CHSH inequalities. In: Khrennikov, A. Yu. (ed.) *Foundations of Probability and Physics - 3*. AIP Conf. Proc., Vol. 750 (American Institute of Physics, Melville, NY 2005), pp. 42–48

18. Arthur, W. B., Holland, J. H., LeBaron, B., Palmer, R. and Tayler, P.: Asset pricing under endogenous expectations in an artificial stock market. In: Arthur, W. B., Durlauf, S. N., and Lane, D. A. (eds.) *The Economy as an Evolving Complex System II*. (Westview, Boulder, CO 1997), pp. 15–44

19. Ashby, W. R.: Adaptiveness and equilibrium. J. Mental Sci. 86, 478–483 (1989)

20. Aspect, A., Dalibard, J., and Roger, G.: Experimental test of Bell's inequalities using time-varying analyzers, Phys. Rev. Lett. 49, 1804–1807 (1982)

21. Atmanspacher, H. and Primas, H.: Epistemic and ontic quantum realities. In: Khrennikov, A. Yu. (ed.) *Foundations of Probability and Physics - 3* AIP, Conf. Proc., Vol. 750 (American Institute of Physics, Melville, NY 2005), pp. 49–62

22. Baaquie, B. E.: *Quantum Finance: Path Integrals and Hamiltonians for Options and Interest Rates* (Cambridge University Press, Cambridge, 2007)

23. Baaquie, B. E.: Quantum mechanics and option pricing. In: Bruza, P. D., Lawless, W., van Rijsbergen, K., Sofge, D. A., Coecke, B., and Clark, S. (eds.) *Quantum Interaction*, Proc. of QI-2008 (College Publications, London 2008), pp.49–53

24. Bachelier, L.: Ann. Sci. Ecole Normale Supérieure 111–17, 21 (1890)

25. Ballentine, L. E.: The statistical interpretation of quantum mechanics. Rev. Mod. Phys. 42, 358–381 (1989)

26. Ballentine, L. E.: *Quantum Mechanics* (Prentice Hall, Englewood Cliffs, NJ 1989)

27. Ballentine, L. E.: Interpretations of probability and quantum theory. In: Khrennikov, A. Yu. (ed.) *Foundations of Probability and Physics*, Quantum Probability and White Noise Analysis, Vol. 13 (World Scientific, Singapore 2001), pp. 71–84

28. Barnett, W. A. and Serletis, A.: Martingales, nonlinearity, and chaos. J. Econ. Dynam. Control 24, 703 (2000)

29. Barrett, J. A.: *The Quantum Mechanics of Minds and Worlds* (Oxford University Press, Oxford 1999)

30. Bearden, J. N., Wallsten, T. S. and Fox, C. R: Error and subadditivity: a stochastic model of subadditivity (University of Arizona, Department of Management and Policy, Tucson, AZ 2005); www.behavioral-or.org/files/SSTFinal.pdf

31. Bell, J.: *Speakable and Unspeakable in Quantum Mechanics* (Cambridge University Press, Cambridge 1987)

32. Beltrametti, E.G. and Cassinelli, G.: *The Logic of Quantum Mechanics*. Encyclopedia of Mathematics and Its Applications, Vol 15 (Cambridge University Press, Cambridge 1984)

33. Bengtsson I.: MUBs, polytopes, and finite geometries. In: Khrennikov, A. Yu. (eds.) *Foundations of Probability and Physics - 3* AIP Conf. Proc., Vol. 750 (American Institute of Physics, Melville, NY 2005), pp. 63–69

34. Benhabib, J.: *Cycles and Chaos in Economic Equilibrium*. (Princeton University Press, Princeton, NJ 1992)

35. Bernasconi, J. and Gustafson, K.: Contextual quick-learning and generalization by humans and machines. Network: Comput. Neural Syst. 9, 85–106 (1998)

36. Bitbol, M., Darrigol, O. (eds.): *Erwin Schrödinger: Philosophy and the Birth of Quantum Mechanics* (Editions Frontières, Gif-sur-Yvette 1992)

37. Black, F. and Scholes, M.: The pricing of options and corporate liabilities. J. Polit. Econ. 81, 637–654 (1973)

38. Bohm, D.: *Unfolding Meaning* (Routledge and Kegan Paul, London 1987)

39. Bohm, D.: *Wholeness and the Implicate Order* (Routledge and Kegan Paul, London 1987)

40. Bohm, D. and Hiley, B.: *The Undivided Universe: An Ontological Interpretation of Quantum Mechanics* (Routledge and Kegan Paul, London 1993)

41. Bohr, N.: Can quantum-mechanical description of physical reality be considered complete? Phys. Rev. 48, 696–702 (1935)

42. Bohr, N.: *The Philosophical Writings of Niels Bohr*, 3 vols. (Ox Bow Press, Woodbridge, CT, 1987)
43. Brock, W. A. and Sayers, C.: Is business cycle characterized by deterministic chaos? J. Monetary Econ. 22, 71–90 (1988)
44. Bruce, V. and Young, A.: Understanding face recognition. Br. J. Psychol. 77, 305–326 (1986)
45. Bulinski, A. V. and Khrennikov, A.Yu.: Nonclassical total probability formula and quantum interference probabilities. Stat. Prob. Lett. 70, 49–58 (2004)
46. Busch, P., Grabowski M. and Lahti, P. J.: *Operational Quantum Physics* (Springer, Berlin 1995)
47. Busch, P.: Less (precision) is more (information): Quantum information in fuzzy probability theory. In: Khrennikov, A.Yu. (ed.) *Quantum Theory: Reconsideration of Foundations - 2*, Ser. Math. Model., Vol. 10 (Växjö University Press, Växjö 2003), pp. 113–148
48. Busemeyer, J. R., Wang, Z. and Townsend, J. T.: Quantum dynamics of human decision making. J. Math. Psychol. 50, 220–241 (2006)
49. Busemeyer, J. R. and Wang, Z.: Quantum information processing explanation for interactions between inferences and decisions. In: Bruza, P. D., Lawless, W., van Rijsbergen, K., Sofge, D. A. (eds.) *Quantum Interaction*, AAAI Spring Symp., Tech. Rep. SS-07-08, (AAAI Press, Menlo Park, CA 2007), pp. 91–97
50. Busemeyer, J. R., Matthews, M., and Wang, Z.: A quantum information processing explanation of disjunction effects. In: Sun, R. and Myake, N. (eds.) *The 29th Annu. Conf. of the Cognitive Science Society and the 5th Int. Conf. of Cognitive Science* (Erlbaum, Mahwah, NJ 2006), pp. 131–135
51. Busemeyer, J. R., Santuy, E. and Lambert-Mogiliansky, A.: Comparison of Markov and quantum models of decision making. In Bruza P., Lawless W., van Rijsbergen K., Sofge D. A., Coecke B. and Clark S. (eds.) *Quantum Interaction*: Proc. of the 2nd Quantum Interaction Symp. (College Publications, London 2008), pp. 68–74
52. Campbell, J. Y., Lo, A. W. and MacKinlay, A. C.: *The Econometrics of Financial Markets* (Princeton University Press, Princeton, NJ 1997)
53. Choustova, O. A.: Pilot wave quantum model for the stock market. In: Khrennikov, A. Yu. (ed.) *Quantum Theory: Reconsideration of Foundations*, Ser. Math. Model., Vol. 2 (Växjö University Press, Växjö 2002), pp. 41–58
54. Choustova, O.: Bohmian mechanics for financial processes. J. Mod. Opt. 51, 1111 (2004)
55. Choustova, O.: Quantum socio-financial model. In: Proc. 9th World Multi-Conference on Systemics, Cybernetics and Informatics, July 10–13, 2005 Orlando, Vol. 1, 339–343 (2005)
56. Choustova, O.: Quantum Bohmian model for financial market. Physica A 374, 304–314 (2006)
57. Choustova, O.: Price-dynamics of shares and Bohmian mechanics: Deterministic or stochastic model? In: Adenier, G., Fuchs, C. and Khrennikov, A. Yu. (eds.) *Foundations of Probability and Physics - 4*, AIP Conf. Proc., Vol. 889 (American Institute of Physics, Melville, NY 2007), pp. 274–282
58. Choustova, O.: Quantum modeling of nonlinear dynamics of stock prices: Bohmian approach. Theor. Math. Phys. 152, 405–416 (2007)
59. Choustova, O.: Quantum model for the price dynamics: The problem of smoothness of trajectories. J. Math. Anal. Appl. 346, 296–304 (2008)
60. Choustova, O.: Application of Bohmian mechanics to dynamics of prices of shares: Stochastic model of Bohm-Vigier from properties of price trajectories. Int. J. Theor. Phys. 47, 252–260 (2008)
61. Choustova, O.: Quantum probability and financial market. Information Sci. 179, 478–484 (2009)
62. Choustova, O.: Quantum-like viewpoint on the complexity and randomness of the financial market. In: Faggini, M. and Lux, T. (eds.) *Coping with the Complexity of Economics* (Springer, Milan 2009), pp. 53–66
63. Cohen, D.: *An Introduction to Hilbert Space and Quantum Logic* (Springer, New York 1989)

64. Cohen, J. D. , Perlstein, W. M., Braver, T. S., Nystrom, L. E., Noll, D. C., Jonides, J. and Smith, E. E.: Temporal dynamics of brain activation during working memory task. Nature 386, 604–608 (1997)
65. Collishaw, S., and Hole, G.: Featural and configurational processes in the recognition of faces of different familiarity. Perception 29, 893–909 (2000)
66. Conte, E., Todarello, O., Federici, A., Vitiello, F., Lopane, M., Khrennikov, A. and Zbilut, J. P.: Some remarks on an experiment suggesting quantum-like behavior of cognitive entities and formulation of an abstract quantum mechanical formalism to describe cognitive entity and its dynamics. Chaos, Solitons and Fractals 31, 1076–1088 (2006)
67. Conte, E., Khrennikov, A. Yu., Todarello, O., Federici, A., and Zbilut, J. P.: A preliminary experimental verification on the possibility of Bell inequality violation in mental states. Neuroquantology 6, 214–221 (2008)
68. Conte, E., Khrennikov, A., Todarello, O., Federici, A., Mendolicchio, L. and Zbilut, J. P.: Mental states follow quantum mechanics during perception and cognition of ambiguous figures. Open Syst. Inform. Dynam. 16, 1–17 (2009)
69. Courtney, S. M., Ungerleider, L. G., Keil, K. and Haxby J. V.: Transient and sustained activity in a disturbed neural system for human working memory. Nature 386, 608–611 (1997)
70. Cressman, R.: *Evolutionary Dynamics and Extensive Form Games* (MIT Press, Cambridge, MA 2004)
71. Croson, R.: The disjunction effect and reasoning-based choice in games. Org. Behavior Human Decision Process. 80, 118–133 (1999)
72. Cushing, J., Fine, A. and Goldstein, S. (eds.): *Bohmian Mechanics and Quantum Theory – An Appraisal* (Kluwer Academic, Dordrecht 1996)
73. Danilov, V. I. and Lambert-Mogiliansky, A.: Non-classical expected utility theory with application to type indeterminacy. PSE Working papers 2007-36 (Paris-Jourdan Sciences Economiques, Paris 2007)
74. Danilov, V. I. and Lambert-Mogiliansky, A.: Non-classical measurement theory: a framework for behavioral sciences. PSE Working papers 2005-37 (Paris-Jourdan Sciences Economiques, Paris 2005); arXiv:physics/0604051 (2006)
75. Danilov, V. I. and Lambert-Mogiliansky, A.: Measurable systems and behavioral sciences. Math. Social Sci. 55, 315–340 (2008)
76. Danilov, V. I. and Lambert-Mogiliansky, A.: Decision making under nonclassical uncertainty. In: Bruza, P. D., Lawless, W., van Rijsbergen, K., Sofge, D. A., Coecke, B., and Clark, S. (eds.) *Quantum Interaction*, Proc. QI-2008 (College Publications, London 2008), pp. 83–86.
77. D'Ariano, G. M.: Operational axioms for quantum mechanics. In: Adenier, G., Fuchs, C. and Khrennikov, A. Yu. (eds.) *Foundations of Probability and Physics - 4*, AIP Conf. Proc., Vol. 889 (American Institute of Physics, Melville, NY 2007), pp. 79–105
78. De Baere, W.: Einstein–Podolsky–Rosen paradox and Bell's inequalities. Adv. Electron. Electron Phys. 68, 245–336 (1986)
79. De Baere, W.: Subquantum nonreproducibility and the complete local description of physical reality. In: Khrennikov, A. Yu. (ed.) *Quantum Theory: Reconsideration of Foundations*. Ser. Math. Model., Vol. 2 (Växjö University Press, Växjö 2002), pp. 59–74
80. de Broglie, L.: *The Current Interpretation of Wave Mechanics: A Critical Study* (Elsevier, Amsterdam 1964)
81. De Coster, G. P. and Mitchell, D. W.: Nonlinear monetary dynamics. J. Business Econ. Stat. 9, 455–462 (1991)
82. de la Peña, L. and Cetto, A. M.: *The Quantum Dice: An Introduction to Stochastic Electrodynamics* (Kluwer Academic, Dordrecht 1996)
83. de la Peña, L. and Cetto, A. M.: Recent developments in linear stochastic electrodynamics. In: Adenier, G., Khrennikov, A. Yu. and Nieuwenhuizen, T. M. (eds.) *Quantum Theory: Reconsideration of Foundations - 3*, AIP Conf. Proc., Vol. 810 (American Institute of Physics, Melville, NY 2006), pp. 131–140

84. de Muynck, W. M.: Interpretations of quantum mechanics, and interpretations of violations of Bell's inequality. In: Khrennikov, A.Yu. (ed.) *Foundations of Probability and Physics*, Qp-Pq: Quantum Probability and White Noise Analysis, Vol. 13 (World Scientific, Singapore 2001), pp. 95–104

85. de Muynck, W. M.: *Foundations of Quantum Mechanics, an Empiricist Approach* (Kluwer Academic, Dordrecht 2002)

86. Deutsch, D.: *The Fabric of Reality* (Allen Lane, London 1997)

87. d'Espagnat, B.: *Veiled Reality: An Analysis of Present-Day Quantum Mechanical Concepts* (Addison-Wesley, Reading, MA 1995)

88. d'Espagnat, B.: *Conceptual Foundations of Quantum Mechanics* (Perseus Books, Reading, MA 1999)

89. Diamond, R. and Carey, S.: Why faces are and are not special: an effect of expertise. J. Experimental Psychol.: Gen. 115, 107–117 (1986)

90. Dirac, P. A. M.: *The Principles of Quantum Mechanics* (Clarendon Press, Oxford 1995)

91. Donald, M. J.: On many-minds interpretations of quantum theory. arXiv:quant-ph/9703008v2 (1997)

92. Donald, M. J.: Quantum theory and the brain. Proc. R. Soc. London. A 427, 43–93 (1990)

93. Donald, M. J.: A mathematical characterization of the physical structure of observers. Found. Phys. 25, 529–571 (1996)

94. Dubois, D. and Prade, H.: Modelling and inductive inference: a survey of recent non-additive probability systems. Acta Psychol. 68, 53–78 (1988)

95. Dvurečenskij, A. and Pulmannová, S.: *New Trends in Quantum Structures* (Kluwer Academic, Dordrecht 2000)

96. Eccles, J. C.: *The Understanding of the Brain* (McGraw-Hill, New York 1974)

97. Edelman, G. M.: *The Remembered Present: A Biological Theory of Consciousness* (Basic Books, New York 1974)

98. Edwards, S. F. and Anderson, P. W.: Theory of spin glasses. J. Phys. F 5, 965 (1975)

99. Einstein, A., Podolsky, B., Rosen, N.: Can quantum-mechanical description of physical reality be considered complete? Phys. Rev. 47, 777–780 (1935)

100. Ekert, A. K.: Quantum cryptography based on Bell's theorem. Phys Rev. Lett. 67, 661–663 (1999)

101. Ezhov, A. A. and Khrennikov, A. Yu.: Agents with left and right dominant hemispheres and quantum statistics. Phys. Rev. E 71, 016138-1–8 (2005)

102. Ezhov, A. A., Khrennikov, A. Yu. and Terentyeva, S. S.: Indications of a possible symmetry and its breaking in a many-agent model obeying quantum statistics. Phys. Rev. E 77, 031126-1–12 (2008)

103. Fama, E. F.: Efficient captial markets: a review of theory and empirical work. J. Finance 25, 383 (1970)

104. Feynman, R. P. and Hibbs, A.: *Quantum Mechanics and Path Integrals* (McGraw-Hill, New York 1965)

105. Feynman, R. P.: Negative probability. In: Hiley, B. J. and Peat, F. D. (eds.) *Quantum Implications, Essays in Honour of David Bohm* (Routledge and Kegan Paul, London 1987), pp. 235–246

106. Folse, H. J.: Bohr's conception of the quantum mechanical state of a system and its role in the framework of complementarity. In: Khrennikov, A. Yu. (ed.) *Quantum Theory: Reconsideration of Foundations*, Ser. Math. Model., Vol. 2 (Växjö University Press, Växjö 2002), pp. 83–98

107. Fox, C., Rogers, B. and Tversky, A.: Option traders exhibit subadditive decision weights. J. Risk Uncertainty 13, 5–17 (1996)

108. Foulis, D. J.: A half-century of quantum-logic. What have we learned? In: Aerts, D. and Pykacz, J. (eds.) *Quantum Structures and the Nature of Reality* (Kluwer Academic, Dordrecht 1990), pp. 1–36

109. Franco, R.: Quantum mechanics, Bayes' theorem and the conjunction fallacy. arXiv:quant-ph/0703222 (2007)
110. Franco, R.: Quantum amplitude amplification algorithm: an explanation of availability bias. arXiv:0810.0959 (2008)
111. Franco, R.: Grover's algorithm and human memory. arXiv:0804.3294 (2008)
112. Franco, R.: Belief revision in quantum decision theory: gambler's and hot hand fallacies. arXiv:0801.4472 (2008)
113. Franco, R.: Risk, ambiguity and quantum decision theory. arXiv:0711.0886 (2007)
114. Franco, R.: The conjunction fallacy and interference effects. J. Math. Psychol. (2009), in press
115. Freud, S.: *New Introductory Lectures on Psychoanalysis* (Norton, New York 1933)
116. Fuchs, C. A.: The anti-Växjö interpretation of quantum mechanics. In: Khrennikov, A. Yu. (ed.) *Quantum Theory: Reconsideration of Foundations*, Ser. Math. Model., Vol. 2 (Växjö University Press, Växjö 2002), pp. 99–116
117. Fuchs, C.: Delirium quantum (or, where I will take quantum mechanics if it will let me). In: Adenier, G., Fuchs, C. and Khrennikov, A. Yu. (eds.) *Foundations of Probability and Physics - 3*, AIP Conf. Proc., Vol. 889 (American Institute of Physics, Melville, NY 2007), pp. 438–462
118. Gacs, P.: On the symmetry of algorithmic information, Sov. Math. – Dokl. 15, 1477–1480 (1974)
119. Gill, R.: Time, finite statistics and Bell's fifth position. In: Khrennikov, A. Yu. (ed.) *Foundations of Probability and Physics - 2*, Ser. Math. Model., Vol. 5 (Växjö University Press, Växjö 2003), pp. 179–206
120. Gisin, N. and Gisin, B.: A local hidden variable model of quantum correlation exploiting the detection loophole. Phys. Lett. A 260, 323–327 (1999)
121. Gnedenko, B. V.: *Theory of Probability*, 6th edn. (Gordon and Breach, London 1998)
122. Granger, C. W. J.: Is chaotic theory relevant for economics? A review essay. J. Int. Comparative Econ. 3, 139–145 (1994)
123. Green, M. B., Schwarz, J. H. and Witten, E.: *Superstring Theory* (Cambridge University Press, Cambridge 1987)
124. Grib, A., Khrennikov, A. Yu. and Starkov, K.: Probability amplitude in quantum-like games. In: Khrennikov, A. Yu. (ed.) *Quantum Theory: Reconsideration of Foundations 2*, Ser. Math. Model., Vol. 10 (Växjö University Press, Växjö 2004), pp. 703–722
125. Grib, A., Khrennikov, A. Yu., Parfionov, G. and Starkov, K.: Quantum equilibria for macroscopic systems. J. Phys. A.: Math. Gen. 39, 8461–8475 (2006)
126. Gudder, S. P.: *Axiomatic Quantum Mechanics and Generalized Probability Theory* (Academic, New York 1970)
127. Gudder, S. P.: An approach to quantum probability. In: Khrennikov, A. Yu. (ed.) *Foundations of Probability and Physics*, Qp-Pq: Quantum Probability and White Noise Analysis, Vol. 13 (World Scientific, Singapore 2001), pp. 147–160
128. Hameroff, S.: Quantum coherence in microtubules. A neural basis for emergent consciousness? J. Consciousness Stud. 1, 91–118 (1994)
129. Hameroff, S.: Quantum computing in brain microtubules? The Penrose–Hameroff "Orch OR" model of consciousness. Philos. Trans. R. Soc. London A 356, 1869–1896 (1998)
130. Hardy, L.: Quantum theory from intuitively reasonable axioms. In: Khrennikov, A. Yu. (ed.) *Quantum Theory: Reconsideration of Foundations*, Ser. Math. Model. Vol. 2 (Växjö University Press, Växjö 2002), pp. 117–130
131. Harkavy, A. A.: Quantum mechanical information is ubiquitous. In: Kafatos, M. (ed.) *Bell's Theorem, Quantum Theory and Conceptions of the Universe* (Kluwer Academic, Dordrecht 1989), pp. 11–24
132. Haven, E.: A discussion on embedding the Black–Scholes option pricing model in a quantum physics setting. Physica A 304, 507–524 (2002)

133. Haven, E.: A Black–Scholes Schrödinger option price: 'bit' versus 'qubit'. Physica A 324, 201–206 (2003)
134. Haven, E.: The wave-equivalent of the Black–Scholes option price: an interpretation. Physica A 344, 142–145 (2004)
135. Haven, E.: Analytical solutions to the backward Kolmogorov PDE via an adiabatic approximation to the Schrödinger PDE. J. Math. Anal. Appl. 311, 439–444 (2005)
136. Haven, E.: Bohmian mechanics in a macroscopic quantum system, In: Adenier, G., Khrennikov, A. Yu. and Nieuwenhuizen, T. M (eds.) *Quantum Theory: Reconsideration of Foundations - 3*, AIP Conf. Proc., Vol. 810 (American Institute of Physics, Melville, NY 2006), pp. 330–340
137. Haven, E. and Khrennikov, A. Yu.: Quantum mechanics and violation of the sure-thing perinciple: the use of probability interference and other concepts. J. Math. Psychol. (2009), in press
138. Healey, R.: How many worlds? Nous 18, 591–616 (1984)
139. Heisenberg, W.: *The Physical Principles of the Quantum Theory* (Dover, London 1989)
140. Heisenberg, W.: *Physics and Philosophy* (Harper and Row, New York 1958)
141. Helstrom, C. W.: *Quantum Detection and Estimation Theory* (Academic, New York 1976)
142. Hess, K. and Philipp, W.: A possible loophole in the theorem of Bell. Proc. Natl. Acad. Sci. USA 98, 14224–14227 (2001)
143. Hess, K. and Philipp, W.: Bell's theorem: critique of proofs with and without inequalities. In: Khrennikov, A. Yu. (ed.): *Foundations of Probability and Physics - 3*, AIP Conf. Proc., Vol. 750 (American Institute of Physics, Melville, NY 2005), pp. 150–155
144. Hiley, B. and Pylkkänen, P.: Active information and cognitive science – A reply to Kieseppä. In: Pylkkänen, P., Pylkkö, P., Hautamäki, A. (eds.) *Brain, Mind and Physics* (IOS Press, Amsterdam 1997), pp. 123–145
145. Hofstadter, D. R.: Dilemmas for superrational thinkers, leading up to a luring lottery. Sci. Am. 6 (1983, June)
146. Hofstadter, D. R.: *Metamagical Themes: Questing for the Essence of Mind and Pattern* (Basic Books, New York 1985)
147. Holevo, A. S.: Investigations on general theory of statistical decisions. Proc. Steklov Inst. Math. 3, 1–180 (1973)
148. Holevo, A. S.: *Statistical Structure of Quantum Theory* (Springer, Berlin 2001)
149. Holland, P.: *The Quantum Theory of Motion* (Cambridge University Press, Cambridge 1993)
150. Hopfield, J. J.: Neural networks and physical systems with emergent collective computational abilities. Proc. Natl. Acad. Sci. USA 79, 1554–2558 (1982)
151. Hoppensteadt, F. C.: *An Introduction to the Mathematics of Neurons: Modeling in the Frequency Domain* (Cambridge University Press, New York 1997)
152. Hsieh, D. A.: Chaos and nonlinear dynamics: application to financial markets. J. Finance 46, 1839–1877 (1991)
153. Jahn, R. G. and Dunne, B. J.: On the quantum mechanics of consciousness, with applications to anomalous phenomena. Found. Phys. 16, 721–772 (1986)
154. Jibu, M. and Yasue, K.: *Quantum Brain Dynamics and Consciousness* (John Benjamins, Amsterdam 1984)
155. Jung, C. G., Pauli, W. and Meier, C. A. (ed.): *Ein Briefwechsel 1932–1958* (Springer, Berlin 1992), pp. 679–702
156. Jung, C. G. and Pauli, W.: *Naturerklärung und Psyche* (Rascher, Zurich 1952)
157. Khrennikov, A. Yu.: *p*-adic quantum mechanics with *p*-adic valued functions. J. Math. Phys. 32, 932–937 (1984)
158. Khrennikov, A. Yu.: *p-Adic Valued Distributions in Mathematical Physics* (Kluwer Academic, Dordrecht 1994)
159. Khrennikov, A. Yu.: *Non-Archimedean Analysis: Quantum Paradoxes, Dynamical Systems and Biological Models* (Kluwer Academic, Dordrecht 1997)

160. Khrennikov, A. Yu.: *Superanalysis* (Nauka, Fizmatlit, Moscow 1997), in Russian; (2nd edn. Nauka, Fizmatlit, Moscow 2005); English translation: *Superanalysis* (Kluwer Academic, Dordrecht 1999)

161. Khrennikov, A. Yu.: *Interpretations of Probability* (VSP International Science Publishers, Utrecht 1999); 2nd enlarged edn. (de Gruyter, Berlin 2009)

162. Khrennikov, A. Yu.: Statistical measure of ensemble nonreproducibility and correction to Bell's inequality. Nuovo Cimento B 115, 179–184 (1999)

163. Khrennikov, A. Yu.: Linear representations of probabilistic transformations induced by context transitions. J. Phys. A: Math. Gen. 34, 9965–9981 (2001)

164. Khrennikov, A. Yu.: Frequency analysis of the EPR-Bell argumentation. Found. Phys. 32, 1159–1174 (2002)

165. Khrennikov, A. Yu. (ed.): *Quantum Theory: Reconsideration of Foundations*, Ser. Math. Model., Vol. 2 (Växjö University Press, Växjö 2002)

166. Khrennikov, A. Yu. and Volovich, I. V.: Quantum nonlocality, EPR model, and Bell's theorem. In: Semikhatov, A., Vasiliv, M., Zaikin, V. (eds.) *3rd International Sakharov Conference on Physics: Proceedings*, Vol. 2 (Scientific World, Moscow 2003), pp. 269–276

167. Khrennikov, A. Yu. (ed.): *Foundations of Probability and Physics - 2*, Ser. Math. Model., Vol. 5 (Växjö University Press, Växjö 2003)

168. Khrennikov, A.Yu. and Smolyanov, O. G.: Nonclassical Kolmogorovian type models describing quantum experiments. Dokl. Akad. Nauk 388, 27–32 (2003); English translation: Dokl. Math. 67, 93–97 (2003)

169. Khrennikov, A. Yu. and Kozyrev, S. V.: Noncommutative probability in classical disordered systems. Physica A 326, 456–463 (2003)

170. Khrennikov, A. Yu.: Contextual viewpoint to quantum stochastics. J. Math. Phys. 44, 2471-2478 (2003)

171. Khrennikov, A. Yu.: Representation of the Kolmogorov model having all distinguishing features of quantum probabilistic model. Phys. Lett. A 316, 279–296 (2003)

172. Khrennikov, A. Yu.: Hyperbolic quantum mechanics Adv. Appl. Clifford Algebras 13, 1–9 (2003)

173. Khrennikov, A. Yu.: Quantum-like formalism for cognitive measurements. Biosystems 70, 211–233 (2003)

174. Khrennikov, A. Yu.: Interference of probabilities and number field structure of quantum models. Annalen der Physik 12, 575–585 (2003)

175. Khrennikov, A. Yu.: Quantum-psychological model of the stock market. Problems and Perspectives in Management 1, 137–148 (2003)

176. Khrennikov, A. Yu.: *Information Dynamics in Cognitive, Psychological, Social, and Anomalous Phenomena* (Kluwer Academic, Dordrecht 2004)

177. Khrennikov, A. Yu.: Växjö interpretation-2003: Realism of contexts. In: Khrennikov, A. Yu. (ed.) *Quantum Theory: Reconsideration of Foundations - 2*, Ser. Math. Model., Vol. 10 (Växjö University Press, Växjö 2004), pp. 323–338

178. Khrennikov, A. Yu.: Contextual approach to quantum mechanics and the theory of the fundamental prespace. J. Math. Phys. 45, 902–921 (2004)

179. Khrennikov, A. Yu.: EPR-Bohm experiment and interference of probabilities. Found. Phys. Lett. 17, 691–700 (2004)

180. Khrennikov, A. Yu.:, On quantum-like probabilistic structure of mental information. Open Syst. Information Dynam. 11 (3), 267–275 (2004)

181. Khrennikov, A. Yu.: *Modeling of Processes of Thinking in p-Adic Coordinates* (Nauka, Fizmatlit, Moscow 2004) (in Russian)

182. Khrennikov, A. Yu. (ed.): *Quantum Theory: Reconsideration of Foundations - 2*, Ser. Math. Model., Vol. 10 (Växjö University Press, Växjö 2004)

183. Khrennikov, A. Yu. (ed.): *Foundations of Probability and Physics - 3*. AIP Conf. Proc., Vol. 750 (American Institute of Physics, Melville, NY 2005)

184. Khrennikov, A. Yu. and Smolyanov, O. G.: Probabilistic measurement models with non-commuting and commuting observables. Dokl. Akad. Nauk 402, 748–753 (2005). English Translation: Dokl. Math. 71, 461–465 (2005)

185. Khrennikov, A. Yu.: The principle of supplementarity: a contextual probabilistic viewpoint to complementarity, the interference of probabilities, and the incompatibility of variables in quantum mechanics. Found. Phys. 35, 1655–1693 (2005)

186. Khrennikov, A. Yu.: Interference in the classical probabilistic model and its representation in complex Hilbert space. Physica E 29, 226–236 (2005)

187. Khrennikov, A. Yu.: Interference in the classical probabilistic framework. Fuzzy Sets Syst. 155, 4–17 (2005)

188. Khrennikov, A. Yu.: Hyperbolic quantum mechanics. Dokl. Akad. Nauk 402, 170–172 (2005). English Translation: Dokl. Math. 71, 363–365 (2005)

189. Khrennikov, A. Yu., Smolyanov, O. G., and Truman, A.: Kolmogorov probability spaces describing Accardi models for quantum correlations. Open Syst. Information Dynam. 12, 371–384 (2005)

190. Khrennikov, A. Yu.: On the representation of contextual probabilistic dynamics in the complex Hilbert space: linear and nonlinear evolutions, Schrödinger dynamics. Nuovo Cimento 120, 353–366 (2005)

191. Khrennikov, A. Yu.: A pre-quantum classical statistical model with infinite-dimensional phase space. J. Phys. A: Math. Gen. 38, 9051–9073 (2005)

192. Khrennikov, A. Yu. and Kozyrev, S. V.: Contextual quantization and the principle of complementarity of probabilities. Open Syst. Information Dynam. 12, 303–318 (2005)

193. Khrennikov, A. Yu. and Volovich, I. V.: Local realistic representation for correlations in the original EPR-model for position and momentum. Soft Comput. 10, 521–529 (2005)

194. Khrennikov, A. Yu.: Schrödinger dynamics as the Hilbert space projection of a realistic contextual probabilistic dynamics. Europhys. Lett. 69 (5), 678–684 (2005)

195. Khrennikov, A. Yu.: From classical statistical model to quantum model through ignorance of information. In: Petitjean, M. (ed.) *Proceedings of FIS2005, The Third Conference on the Foundations of Information Science* (MDPI, Basel, Switzerland 2005)

196. Khrennikov, A. Yu.: To quantum mechanics through projection of classical statistical mechanics on prespace. In: Buccheri, R., Elitzur, A. C. and Saniga, M. (eds.) *Endophysics, Time, Quantum and the Subjective* (World Scientific, Singapore 2005), pp. 389–408

197. Khrennikov, A. Yu.: Representation of the contextual statistical model by hyperbolic amplitudes. J. Math. Phys. 46, 062111 (2005)

198. Khrennikov, A. Yu.: Quantum-like brain: "Interference of minds." BioSystems 84, 225–241 (2006)

199. Khrennikov, A. Yu.: Nonlinear Schrödinger equations from prequantum classical statistical field theory. Phys. Lett. A 357, 171–176 (2006)

200. Khrennikov, A. Yu.: Prequantum classical statistical field theory: Complex representation, Hamilton–Schrödinger equation, and interpretation of stationary states. Found. Phys. Lett. 19, 299–319 (2006)

201. Khrennikov, A. Yu.: A formula of total probability with the interference term and the Hilbert space representation of the contextual Kolmogorovian model. Theory Prob. Appl. 51, 427–441 (2007)

202. Khrennikov, A. Yu.: A mathematician's viewpoint to Bell's theorem: in memory of Walter Philipp. In: Adenier, G., Fuchs, C. and Khrennikov, A. Yu. (eds.) *Foundations of Probability and Physics - 3*, AIP Ser. Conf. Proc., Vol. 889 (American Institute of Physics, Melville, NY 2007), pp. 7–17

203. Khrennikov, A. Yu.: Quantumlike representation of extensive form games: Probabilistic aspects. J. Math. Phys. 48, 072107 (2007)

204. Khrennikov, A. Yu.: EPR-Bohm experiment and Bell's inequality: Quantum physics meets probability theory. Theor. Math. Phys. 157, 1448–1460 (2008)

205. Khrennikov, A. Yu.: Quantum-like microeconomics: Statistical model of distribution of investments and production. Physica A 387 5826–5843 (2008)
206. Khrennikov, A. Yu.: Can the mathematical formalism of quantum mechanics be applied to psychology? Proc. SPIE 7023, 702308 (2008)
207. Khrennikov, A. Yu. and Haven, E.: Does probability interference exist in social science? In: Adenier, G., Fuchs, C. and Khrennikov, A. Yu. (eds.), *Foundations of Probability and Physics - 4*, AIP Conf. Proc., Vol. 889 (American Institute of Physics, Melville, NY 2005), pp. 299–309
208. Khrennikov, A. Yu.: Formula of total probability, interference, and quantum-like representation of data for experiments on disjunction effect. In: Bruza, P. D., Lawless, W., van Rijsbergen, K., Sofge, D. A., Coecke, B. and Clark, S. (eds.) *Quantum Interaction*, Proc. QI-2008 (College Publications, London 2008), pp. 34–40
209. Khrennikov, A. Yu.: Classical and quantum-like randomness and the financial market. In: Faggini, M. and Lux, T. (eds.), *Coping with the Complexity of Economics* (Springer, Milan 2009)
210. Khrennikov, A. Yu.: Einstein–Podolsky–Rosen paradox, Bell's inequality, and the projection postulate. J. Russian Laser Res. 29, 101–113 (2008)
211. Khrennikov, A. Yu.: The quantum-like brain on the cognitive and subcognitive time scales. J. Consciousness Stud. 15(7), 39–77 (2008)
212. Khrennikov, A. Yu.: The role of von Neumann and Lüders postulates in the Einstein, Podolsky, and Rosen considerations: comparing measurements with degenerate and nondegenerate spectra. J. Math. Phys. 49, 052102 (2008)
213. Khrennikov, A. Yu.: Quantum-like model of cognitive decision making and information processing. Biosystems 95, 179–187 (2009)
214. Khrennikov, A. Yu.: *Contextual Approach to Quantum Formalism*. Fundamental Theories of Physics, Vol. 160 (Springer, Dordrecht 2009)
215. Kirkpatrick, K. A.: "Quantal" behavior in classical probability. Found. Phys. Lett. 16, 199–224 (2003)
216. Klyshko, D. N.: The Bell and GHZ theorems: a possible three-photon interference experiment and the question of nonlocality, Phys. Lett. A 172, 399–403 (1993)
217. Klyshko, D. N.: Basic quantum mechanical concepts from the operational viewpoint. Usp. Fiz. Nauk 168, 975–1015 (1998)
218. Kochen, S. and Specker, E.: The problem of hidden variables in quantum mechanical systems. J. Math. Mech. 17, 59–87 (1967)
219. Kolmogoroff, A. N.: *Grundbegriffe der Wahrscheinlichkeitsrechnung* (Springer, Berlin 1933); English translation: Kolmogorov, A. N.: *Foundations of the Theory of Probability*, 2nd edn. (Chelsea, New York 1956)
220. Kolmogorov, A. N. and Fomin, S. V.: *Introductory Real Analysis* (Dover, New York 1975)
221. Kunstatter, G., Moffat, J. W. and Malzan, J.: Geometrical interpretation of a generalized theory of gravitation. J. Math. Phys. 24, 886–894 (1983)
222. La Mura, P., Swiatczak, L.: Markovian entanglement networks. arXiv:quant-ph/0702072 (2007)
223. La Mura, P.: Projective expected utility. In: Bruza, P. D., Lawless, W., van Rijsbergen, K., Sofge, D. A., Coecke, B. and Clark, S. (eds.) *Quantum Interaction*, Proc. QI-2008 (College Publications, London 2008), pp. 87–93
224. Lander, K., Bruce, V. and Hill, H.: Evaluating the effectiveness of pixelation and blurring on masking the identity of familiar faces. Appl. Cogn. Psychol. 15, 101–116 (2001)
225. Larsson, J.-Å.: Bell inequalities for position measurements. In: Khrennikov, A. Yu. (ed.) *Quantum Theory: Reconsideration of Foundations - 2*, Ser. Math. Model., Vol. 10 (Växjö University Press, Växjö 2003), pp. 353–364
226. Larsson, J.-Å. and Gill, R.: Bell's inequality and the coincidence time loophole. In: Khrennikov, A. Yu. (ed.) *Foundations of Probability and Physics - 3*, AIP Conf. Proc., Vol. 750 (American Institute of Physics, Melville, NY 2005), pp. 228–235

227. Li, M., Vitanyi, P.: *An Introduction to Kolmogorov Complexity and Its Applications* (Springer, New York 1993)
228. Light, P. and Butterworth, G. (eds.): *Context and Cognition* (Erlbaum, Hillsdale, NJ 1993)
229. Lockwood, M.: *Mind, Brain and the Quantum: The Compound 'I'* (Blackwell, Oxford 1989)
230. Loewer, B.: Comment on Lockwood. Br. J. Philos. Sci. 47, 229–232 (1996)
231. Löngren, L.: Comput. Inform. Sci. 2, 165–175 (1967)
232. Luczak, A., Bartho, P., Marguet, S. L., Buzsáki, G. and Harris, K. D: Sequential structure of neocortical spontaneous activity in vivo. Proc. Natl. Acad. Sci. USA 104, 347–352 (2007)
233. Lüders, G.: Über die Zustandsänderung durch den Messprozess. Ann. Phys. (Leipzig) 8, 322 (1951)
234. Ludwig, G.: *Foundations of Quantum Mechanics* (Springer, Berlin 1983)
235. Mackey, G. W.: *Mathematical Foundations of Quantum Mechanics* (W. A. Benjamin, New York 1963)
236. Mandelbrot, B. B.: The variation of certain speculative prices. J. Business 36, 394–419 (1963)
237. Mantegna, R. N. and Stanley, H. E.: *Introduction to Econophysics* (Cambridge University Press, Cambridge 2000)
238. Marley, J. D. and Hornstein, J.: Quantum statistical inference. Stat. Sci. 8, 433–457 (1993)
239. Mezard, M., Parisi, G., Virasoro, M.: *Spin-Glass Theory and Beyond* (World Scientific, Singapore 1987)
240. Miller, J.: The sampling distribution of d'. Perception Psychophys. 58, 65–72 (1996)
241. Mori, K.: On the relation between physical and psychological time. In: Hameroff, S. (ed.) *Toward a Science of Consciousness* (University Press, Tucson, AZ 2002), p. 102.
242. Muckenheim, W.: A review of extended probability. Phys. Rep. 133, 338–401 (1986)
243. Narens, L.: A new foundation for support theory (Institute for Mathematical Behavioral Sciences, University of California, Irvine 2004), paper 7; http://repositories.cdlib.org/imbs/7
244. Nelson, E.: *Quantum Fluctuations* (Princeton University Press, Princeton 1985)
245. Ohya, M.: Adaptive dynamics and its applications to chaos and NPC problem. In: Accardi, L., Freudenberg, W. and Ohya, M. (eds.) *Quantum Bio-Informatics*. QP-PQ: Quantum Probability and White Noise Analysis, Vol. 21 (World Scientific, Singapore 2008), pp. 181–216
246. Ohya, M.: *Selected Papers of M. Ohya* (World Scientific, Singapore 2008)
247. Orlov, Yu. F.: The wave logic of consciousness: A hypothesis. Int. J. Theor. Phys. 21, 37–53 (1982)
248. Owen, G.: *Game Theory* (Saunders, Philadelphia, PA 1968)
249. Penrose R.: *The Emperor's New Mind* (Oxford University Press, Oxford 1989)
250. Penrose, R.: *Shadows of the Mind* (Oxford University Press, Oxford 1994)
251. Peres, A.: Unperformed experiments have no results. Am. J. Phys. 46, 745–747 (1978)
252. Piotrowski, E. W. and Sladkowski, J.: Quantum-like approach to financial risk: quantum anthropic principle. Acta Phys. Polonica B 32, 3873–3879 (2001)
253. Piotrowski, E. W. and Sladkowski, J.: Quantum market games. Physica A 312, 208–216 (2002)
254. Piotrowski, E. W., Sladkowski, J. and Syska, J.: Interference of quantum market strategies. Physica A 318, 516–528 (2003)
255. Piotrowski, E. W. and Sladkowski, J.: An invitation to quantum game theory. Int. J. Theor. Phys. 42, 1089–1089 (2003)
256. Piotrowski, E. W.: Fixed point theorem for simple quantum strategies in quantum market games. Physica A 324, 196–200 (2003)
257. Piotrowski, E. W. and Sladkowski, J.: Quantum games in finance. Quantitative Finance 4, C61-C67 (2004)
258. Piotrowski, E. W., Schroeder, M. and Zambrzycka, A.: Quantum extension of European option pricing based on the Ornstein–Uhlenbeck process. Physica A 368, 176–182 (2006)
259. Pitowsky I. : From George Boole to John Bell: The Origins of Bell's Inequalities. In: Kafatos, M. (ed.) *Bell's Theorem, Quantum Theory and Conceptions of the Universe* (Kluwer, Dordrecht 1989), pp. 37–49

260. Pitowsky I.: Range Theorems for Quantum Probability and Entanglement. In: Khrennikov A. (ed.) *Quantum Theory: Reconsideration of Foundations*, Ser. Math. Model., Vol. 2 (Växjö University Press, Växjö 2002), pp. 299–308

261. Plotnitsky, A.: Reading Bohr: Complementarity, Epistemology, Entanglement, and Decoherence. In: Gonis, A. and Turchi, P. E. A. (eds.) *Decoherence and its Implications in Quantum Computation and Information Transfer*. NATO Computer and Systems Sciences, Vol. 182 (IOS Press, Amsterdam 2001), pp. 3–37

262. Plotnitsky, A.: *The Knowable and Unknowable: Modern Science, Nonclassical Thought, and the "Two Cultures"* (University of Michigan Press, Ann Arbor, MI 2002)

263. Plotnitsky, A.: "This is an extremely funny thing, something must be hidden behind that": Quantum waves and quantum probability with Erwin Schrödinger. In: Khrennikov, A. Yu. (ed.) *Foundations of Probability and Physics - 3*, AIP Conf. Proc., Vol. 750 (American Institute of Physics, Melville, NY 2005), pp. 388–408

264. Plotnitsky, A.: *Reading Bohr: Physics and Philosophy* (Springer, Dordrecht 2006)

265. Pylkkänen, P. (ed.): *The Search for Meaning* (Aquarian, Wellingborough, UK 1989)

266. Rapoport, A.: Experiments with n-person social traps 1: prisoner's dilemma, weak prisoner's dilemma, volunteer's dilemma, and largest number. J. Conflict Resolution 32, 457–472 (1988)

267. Renyi, A.: *Probability Theory* (North-Holland, Amsterdam 1970)

268. Rottenstreich, Y. and Tversky, A.: Unpacking, repacking and anchoring: advances in support theory. Psychol. Rev. 104, 406–415 (1997)

269. Samuelson, P.: Proof that properly anticipated prices fluctuate randomly. Ind. Management Rev. 6, 41–49 (1965)

270. Samuelson, P.: Rational theory of warrant pricing. Ind. Management Rev. 6, 13–31 (1965)

271. Savage, L. J.: *The Foundations of Statistics* (Wiley, New York 1954)

272. Schneider, W. and Shiffrin, R. M.: Controlled and automatic human information processing. I. Detection, search, and attention. Psychol. Rev. 84, 1–66 (1977)

273. Schrödinger, E.: Die gegenwärtige Situation in der Quantenmechanik. Naturwissenschaften 23, 807–812; 823–828; 844–849 (1935)

274. Segal, W. and Segal, I. E.: The Black–Scholes pricing formula in the quantum context. Proc. Natl. Acad. Sci. USA 95, 4072–4075 (1998)

275. Shafir, E. and Tversky, A.: Thinking through uncertainty: nonconsequential reasoning and choice. Cognitive Psychol. 24, 449–474 (1992)

276. Shiffrin, R. M. and Schneider, W.: Controlled and automatic information processing: II. Perceptual learning, automatic attending, and a general theory. Psychol. Rev. 84, 127–189 (1977)

277. Shimony, A.: *Search for a Naturalistic World View*, Vols. I, II (Cambridge University Press, Cambridge 1993)

278. Shimony, A.: On mentality, quantum mechanics and the actualization of potentialities. In: Penrose, R. and Longair, M. (eds.) *The Large, the Small and the Human Mind* (Cambridge University Press, Cambridge 1997)

279. Shiryaev, A. N.: Kolmogorov: Life and Creative Activities. Annals Prob. 17, 866–944 (1989)

280. Shiryaev, A. N.: *Probability*, 2nd edn. (Springer, New York 1984)

281. Shiryaev, A. N.: *Essentials of Stochastic Finance: Facts, Models, Theory* (World Scientific, Singapore 1999)

282. Soros, J.: *The Alchemy of Finance. Reading the Mind of the Market* (Wiley, New York 1987)

283. Stapp, H. P.: S-matrix interpretation of quantum theory. Phys. Rev. D 3, 1303 (1971)

284. Stapp, H. P.: *Mind, Matter and Quantum Mechanics* (Springer, Berlin 1993)

285. Strogatz, S. H.: *Nonlinear Dynamics and Chaos with Applications to Physics, Biology, Chemistry, and Engineering* (Addison-Wesley, Reading, MA 1994)

286. Suppes, P.: The measurement of belief. J. R. Statist. Soc. B 36, 160–191 (1974)

287. Svozil, K.: *Quantum Logic* (Springer, Berlin 1998)

288. Svozil, K.: *Randomness and Undecidability in Physics* (World Scientific, Singapore 1993)

289. Svozil, K.: On counterfactuals and contextuality. In: Khrennikov, A. Yu. (ed.), *Foundations of Probability and Physics - 3*, AIP Conf. Proc., Vol. 750 (American Institute of Physics, Melville, NY 2005), pp. 351–360

290. 't Hooft, G.: Quantum gravity as a dissipative deterministic system. arXiv:gr-qc/9903084v3 (1999)

291. 't Hooft, G.: The mathematical basis for deterministic quantum mechanics. arXiv:quant-ph/0604008v2 (2006)

292. 't Hooft, G.: The free-will postulate in quantum mechanics. arXiv:quant-ph/0701097v1 (2007)

293. Tversky, A. and Simonson, I.: Context-dependent preferences. Management Sci. 39, 85–117 (1993)

294. Tversky, A. and Koehler, D.: Support theory: a nonexistential representation of subjective probability. Psychol. Rev. 101, 547–567 (1994)

295. Tversky, A. and Shafir, E.: The disjunction effect in choice under uncertainty. Psychol. Sci. 3, 305–309 (1992)

296. van Gelder, T.: What might cognition be, if not computation? J. Philos. 91, 345–381 (1995)

297. van Gelder, T. and Port, R.: It's about time: overview of the dynamical approach to cognition. In: van Gelder, T. and Port, R. (eds.) *Mind as Motion: Explorations in the Dynamics of Cognition* (MIT, Cambridge, MA 1995), pp. 1–43

298. Vitiello, G.: *My Double Unveiled – The Dissipative Quantum Model of Brain* (Benjamins, Amsterdam 2001)

299. von Mises, R.: *The Mathematical Theory of Probability and Statistics* (Academic, London 1964)

300. von Neumann, J. and Morgenstern, O.: *Theory of Games and Economic Behavior* (Princeton University Press, Princeton, NJ 1947)

301. von Neuman, J.: *Mathematical Foundations of Quantum Mechanics* (Princeton University Press, Princeton, NJ 1955)

302. Vorob'ev, N. N. : Consistent families of measures and their extensions. Theory Prob. Appl. 7, 147–162 (1962)

303. Whitehead A. N.: *Process and Reality: An Essay in Cosmology* (Macmillan, New York 1929)

304. Wigner, E. P.: The problem of measurement. Am. J. Phys. 31, 6–46 (1963)

305. Wright, R.: Generalized urn models. Found. Phys. 20, 881–907 (1991)

Index